William Jasper Nicolls

The Railway Builder

A Handbook

William Jasper Nicolls

The Railway Builder
A Handbook

ISBN/EAN: 9783744678827

Printed in Europe, USA, Canada, Australia, Japan

Cover: Foto ©berggeist007 / pixelio.de

More available books at **www.hansebooks.com**

THE RAILWAY BUILDER

A HANDBOOK
FOR ESTIMATING
THE COST OF
AMERICAN RAILWAY
CONSTRUCTION AND
EQUIPMENT

BY

WILLIAM JASPER NICOLLS

M. Am. Soc. C.E.

AUTHOR OF "THE STORY OF AMERICAN COALS," ETC.

*FIFTH EDITION, RE-
VISED & ENLARGED*

PHILADELPHIA

J. B. LIPPINCOTT COMPANY

LONDON : 6 HENRIETTA STREET, COVENT GARDEN

1897

DEDICATED TO
CARL WALDEMAR BUCHHOLZ,
CHIEF ENGINEER
ERIE RAILROAD

BY

THE AUTHOR.

PREFACE TO THE FIFTH EDITION.

In the present edition the entire work has been carefully revised and brought up to date. It has also received many additions, the page enlarged, and a new form of binding adopted, so as to render the volume suitable both for the library and pocket.

The Author has added fifteen years to his professional life, and has learned many useful facts which, as far as practicable, are set forth in this edition for the benefit of the unprofessional reader; for time has shown that to such the book has been useful.

Capitalists who put their money in American railways, contractors who build them, and the host of practical men operating them, will find in the following pages plain and simple directions for estimating on the first cost or for renewals, while the young engineer will find much that heretofore has been covered with many formulas and tedious analyses.

W. J. N.

PHILADELPHIA, 1897.

PREFACE TO THE FIRST EDITION.

In offering this little volume to the railroad world the Author realizes the fact that other books have been published covering the same ground in detail; and that numerous works by Engineers of known ability and reputation are also in existence, which contain an immense fund of general information and most of the tables necessary for calculation. But they are either in a condensed form, or clothed in formulas and symbols totally unknown to the average railroad man. During a professional career of nine years the Author has had a varied experience, embracing nearly every kind of railway construction, locating, building, and equipping them, and during that period has also been closely identified with

11

works manufacturing railway plant. From notes collected during this time, often from the expressed opinions of prominent railway men, from information obtained in travelling over nearly every railway in the United States and the Canadas, and from the works before mentioned, this book has been prepared, not for the critical *savant*, but for the daily use of practical railroad men, those not conversant with Engineering formulas or manufacturers' processes, to enable them to familiarize themselves with the subject and to assist them in estimating the probable cost of constructing and equipping an American railway. The writer acknowledges himself indebted to Trautwine, Haswell, Jervis, Forney, Barry, Gillespie, Haupt, Knight, and Lorenz.

WM. J. NICOLLS.

New York, May 4, 1878.

CONTENTS.

CHAPTER VII.

CHAPTER VIII.

THE RAILWAY BUILDER.

CHAPTER I.

FIELD OPERATIONS.

BEFORE any idea can be formed regarding the probable cost of a projected line of railway, it is necessary to have as complete a map as possible of the proposed line, showing the nature of the country through which the road is to run, together with all available information collected and carefully noted on the map or field-book. For this purpose a party or CORPS OF ENGINEERS must be equipped and sent into "the field," and consists usually of from eight to a dozen men to each corps—graded and termed as follows:—

Chief of Corps, Transitman, Leveller, Level Rodman, Front Chainman, Back

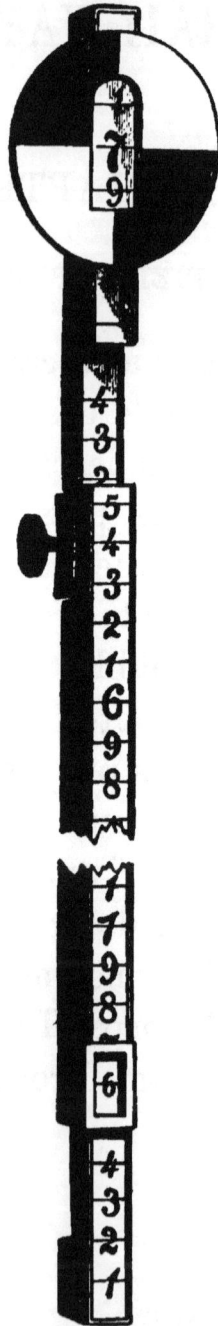

Level rod.

Chainman, Front Rodman, Back Rodman and Axman, to which may be added a Topographer, if necessary. The Chief of Corps is in charge of the party, and his business is to determine what route is to be taken by the line of survey. The Transitman runs the transit instrument, and keeps the "transit field-book." The Leveller runs the level instrument and keeps the "level field-book." Level Rodman handles the level rod, while the two Chainmen and Transit Rodman handle respectively the instruments which their names indicate. The Axman's duties consist in clearing away the brush or trees in advance of the corps, and also in making and driving stakes at the points marked by the Chainman. THE TRANSIT is an instrument used for establishing points in the line of survey and for measuring angles. It consists, essentially, of a circular plate of metal, supported in such a manner as to be horizontal, and divided on its outer circumference into

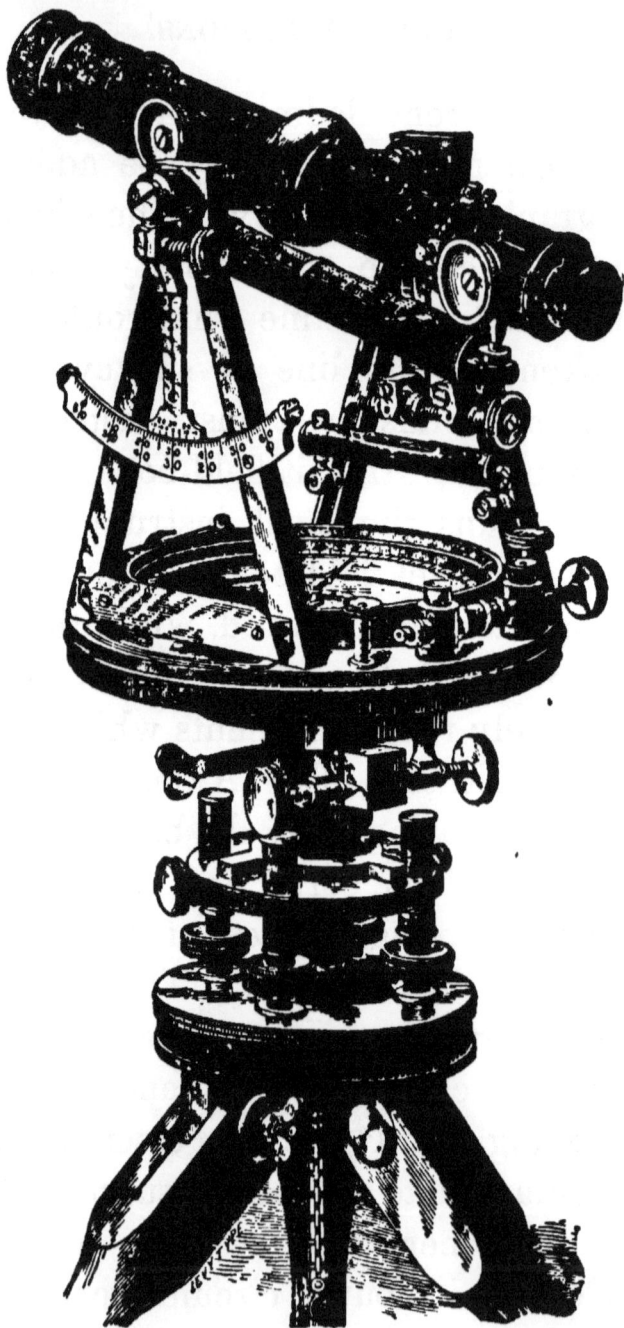

Engineer's transit.

degrees and parts of degrees. Through the centre of this plate passes an upright axis, and on it is fixed a second circular plate, which nearly touches the first plate, and can turn freely around to the right and to the left. This second plate carries a telescope, which rests upon upright standards firmly fixed to the plate, and which can be pointed upwards and downwards. By the combination of this motion, and that of the second plate around its axis, the telescope can be directed to any object. The second plate has some marks on its edge, such as an arrow-head, which serves as a pointer or index for the divided circle, like the hand of a clock. When the telescope is directed to one object, and then turned to the right or to the left to some other object, this index which moves with it and passes around the divided edge of the other plate, points out the one passed over by this change of direction, and thus measures the angles made by the lines imagined to pass from

Engineer's chain.

the centre of the instrument to the two
objects. The cost of one of these instruments, suitable for railroad work, is about
$200. THE ENGINEER'S LEVEL consists
simply of a telescope suspended in the two
arms of a Y, and attached to it is a small
spirit level. By means of screws operated
by the Leveller, the telescope is made level,
and when in this position the Leveller
reads from his rod the height of his instrument and by calculations—explained
hereafter—determines the different elevations of the country through which the
line of survey is passing. This instrument
will cost about $150. THE CHAIN is used
for measuring the length of the lines
which have previously been determined
by the Transitman. It is composed of
100 links of wire, steel or iron, each link
being one foot in length; at every tenth
link is fastened a brass tag, having one,
two, three, or four points, corresponding
to the number of tens which it makes,
counting from the nearest end of the

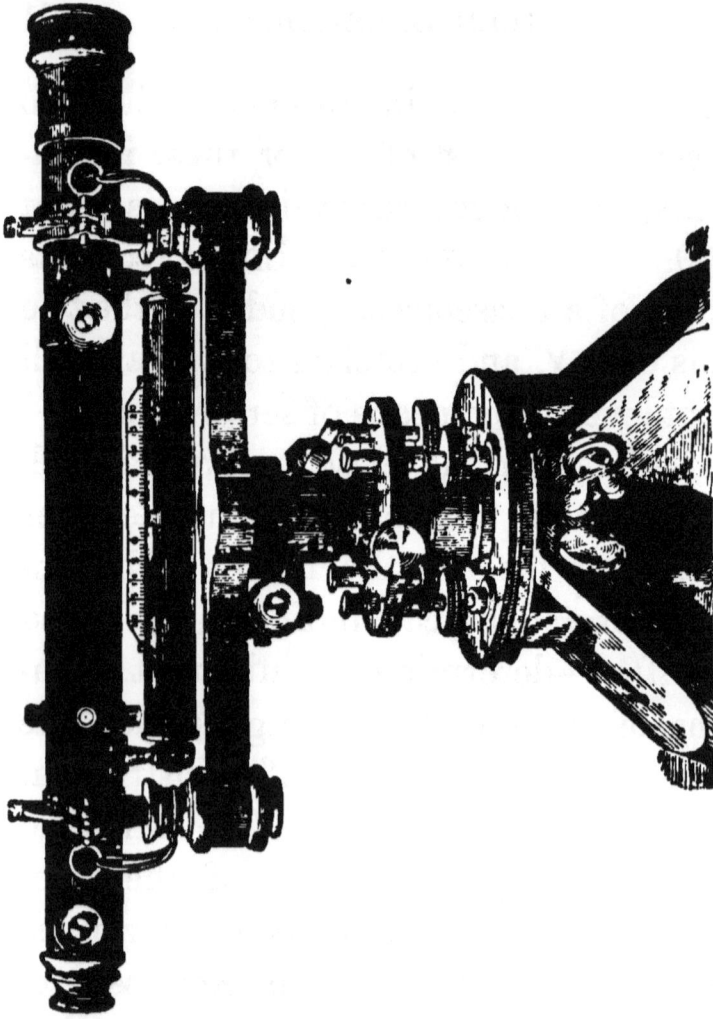

Engineer's level.

chain. The middle of the chain, or fiftieth link is marked by a round piece of brass. A good steel chain 100 feet in length is worth from $10 to $15. An outfit for a field party, sufficiently complete for all practical purposes, would be about as follows:—

1. Engineer's Transit	$200.00
1. " Level	150.00
1. " Level Rod	15.00
1. Steel Chain	15.00
2. Transit Rods @ $1.00 . . .	2.00
1. 100 ft. tape, $5, 1–50 ft. tape, $3 .	8.00
2. Short-handled Axes @ $1.50 . .	3.00
1. Pocket Level	8.00
1. Slope "	6.00
Incidentals	10.00
	$417.00

In running a preliminary line for a railroad, straight lines only are located on the ground, and where these straight lines intersect or join each other, an angle is formed, the size of which is exactly determined by the transit, so named, because the telescope of the instrument is capable

Pocket level.

24

of making a complete revolution on its
axis, which is not the case with the theo-
dolite. In running a line with the transit,
we first ascertain at what point the survey
is to commence, then we designate that
point as station O. Now set the transit
exactly over this point by means of the
"plumb bob" and line which is suspended
from it, and level it up. Direct the telescope to the rod, which is held by the
"front rodman" at the point determined
upon by the "Chief." Sight to the lowest
visible point of the rod, clamp the instrument, and when the needle of the compass
comes to a rest, read the course or bearing of the line which connects these two
points. The bearing is noted by reading
between what letters on the compass the
end of the needle comes, and to what
number, naming or writing down FIRSTLY,
the letter N. or S. (North or South), which
is at the 0° point nearest to the end of
the needle from which you are reading.
SECONDLY, the number of degrees to

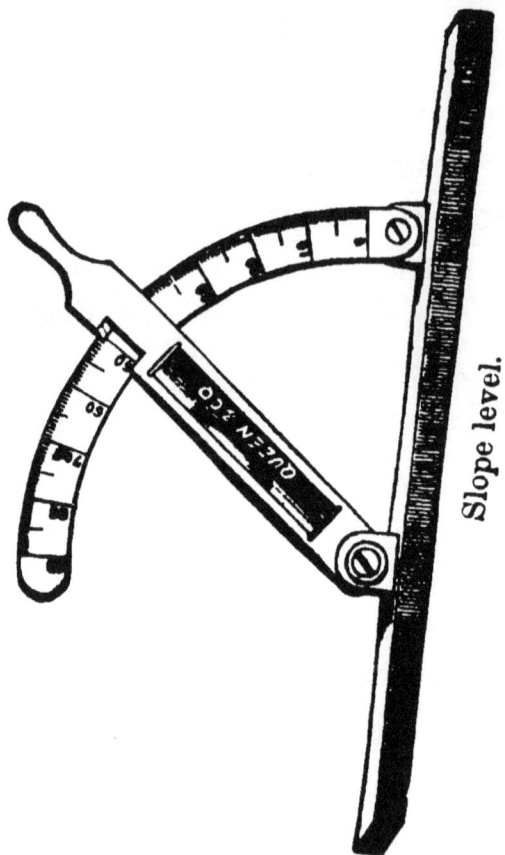

Slope level.

QUEEN & CO.

26

which it points; and THIRDLY, the letters E. or W. (East or West), of the 90° point which is nearest to the same end of the needle.

After this course is carefully noted in the field-book (suppose it to be N. 45° E.), direct the chainmen to measure the distance from the transit point to the point at which the Front Rodman is stationed, the axman driving a stake every hundred feet (1 chain), and exactly in the line given him by the Transit-man with his instrument, after which take a second sight on the front rod to see that the instrument has not been moved in setting the stakes, and also take a second reading of the compass to avoid any error. The Transitman, after directing his Back Rodman to occupy the position at station O, will then move up the instrument to the point previously established by the Front Rodman, where he will set up the transit, as before. The distance from station O to this point has now been ascertained by the Chainmen, and a stake has been driven every hundred feet by the Axman, and numbered 1, 2, 3, 4, etc. Suppose the distance measured to be 800 feet, or 8 chains, then the new point will be at station 8. The Transitman notes this distance in his field-book, and after adjusting his instrument so that the VERNIER plate indicates 0°, the telescope is reversed, and he takes a "back sight" on the Back Rodman who is holding his rod on the head of a tack which is driven in the stake marked O. After clamping the instrument the telescope is then reversed and directed to the Front Rodman, who has again taken up a position in advance which is indicated to him by the Chief as the point to which the line is to be run. At this point there will be an angle, read it on the VERNIER PLATE, and also note the new reading given by the needle, taking care to observe whether the change of direction is to the left or to the right.

The reading of the VERNIER (so named from its inventor, Pierre Vernier, who gave a description of it in a tract published at Brussels in 1631) is the stumbling-block in the use of the transit by many practical men. It is very simple. The outside circle or plate is divided ' into degrees and half degrees, and the inside one into minutes. Now, the vernier is constructed in this way: A length on the circumference is made on the inner plate equal to *twenty-nine* half degrees of the outside plate. This length on the circumference of the inner plate is then divided into *thirty* equal parts, or one more than the number of half degrees occupying the same space on the outer plate. It is obvious, therefore, that each division on the inner plate is a trifle smaller than a half degree on the outer plate, and this trifling difference is the space measured by the vernier.

To READ THE VERNIER.—First, note

the position of 0° on the *inner* plate (usually indicated by an arrow-head), and if it is exactly in line with any division of degrees or half degrees on the outer plate, that will be the angle measured by the vernier. But, if the 0° (or arrow-line) does not coincide exactly with any line on the outer plate, then observe which two lines on the outer and inner plates coincide, or together form a straight line; this, plus the nearest reading indicated on the outside plate by the pointer or arrow-line, will be the correct angle in degrees and minutes.

If several lines seem to coincide, take the middle one.

A brief study of the following graphic figures from Gillespie's "Land Surveying" will make the above description of the vernier more plain. In the first, or vernier A, the reading is 0°, or 360° as indicated by the arrow-line. In the second, or vernier B, the dotted and

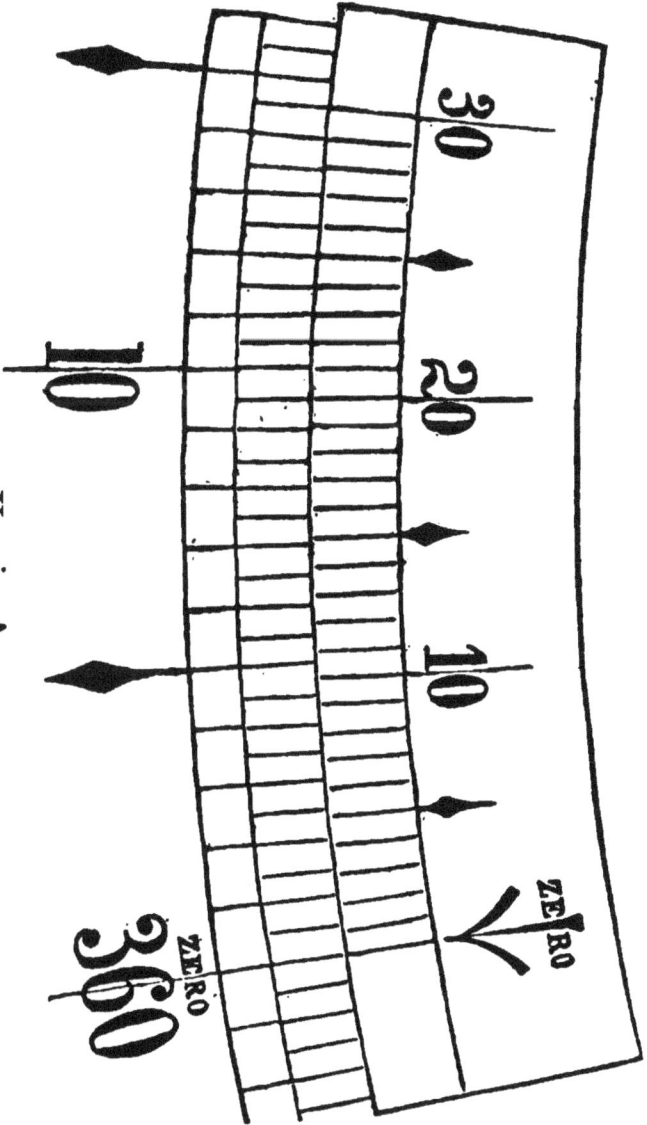

Vernier A.

crossed line shows what divisions coincide, and the reading is 20° 10′, the 0°
(or arrow-line) of the inner plate being
at a point of the outer circle 10′ (minutes) beyond 20° (degrees).

Sometimes the graduations of a vernier are made so as to read both ways
from the arrow-line, or zero. In that
case the vernier is double. Care must
be taken when using this style of vernier to note which way the angle is
measured; that is, if from the arrow-
line, or zero, towards the right hand,
then the reading must be made from
the right-hand half of the vernier.
But if the angle is measured from the
arrow-line, or zero, towards the left
hand, then the reading must be made
from the left-hand half of the vernier.

To return to our survey, suppose the
angle we have just turned with the instrument should read on the vernier 15°
30′ to the right, note it in your field-book
and direct the Chainmen and Axmen

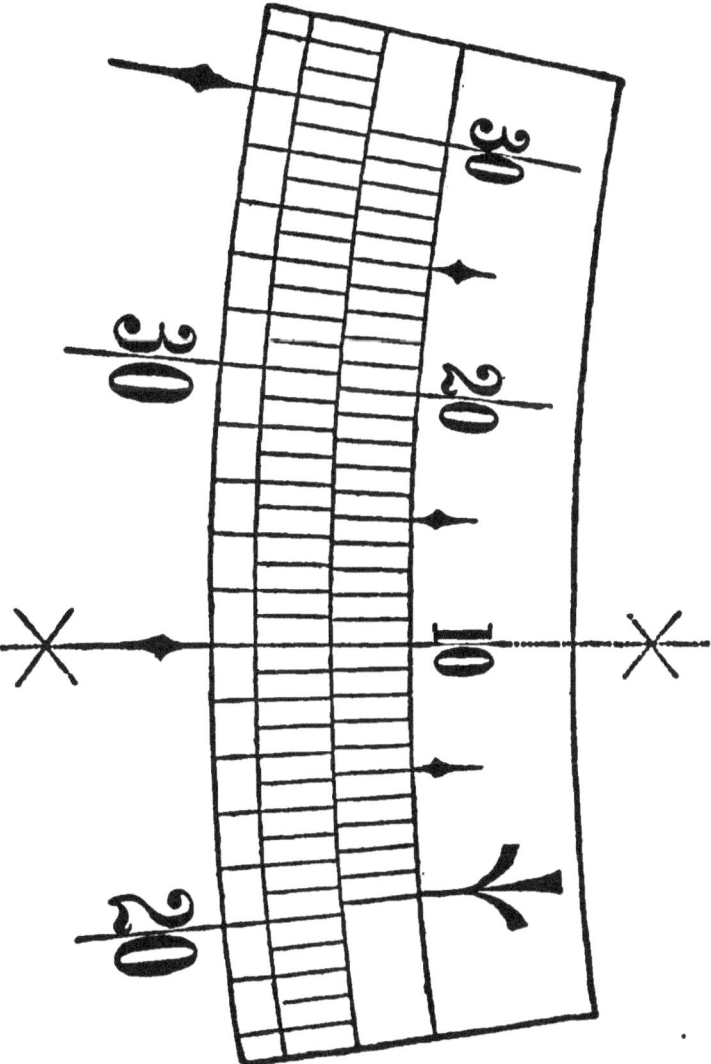

Vernier B.

to proceed as before, and so continue. The field-book will then look as follows:

TRANSIT BOOK.

Sta- tion.	Dis- tance	Angle.	Course.	Remarks.
10				
9				
8⊙	—	15° 30′ R.	N.60° 30′ E.	Plug
7				
6	800′			
5				
4				
3				
2				
1				
0⊙			N. 45° E.	

Fig. 1.

0. On plug intersection of centre line of Madison and Elm Streets.

And so proceed, noting carefully the topography of the country through which the line is running—in the field-book—and marking the distances to any prominent object from the line, such as a large tree, house, barn, stream or river. This can be done by the Transitman, and noted in his field-book as above, but it is much preferable to have an extra man to take the topography of the country in a separate book, thus avoiding confusion of figures, and giving the Transitman more time to devote to his calculations and instrument. In running a " line of survey" it frequently occurs that obstacles to measurement are met which will necessitate a knowledge of triangulation. The simplest forms are noted below :—

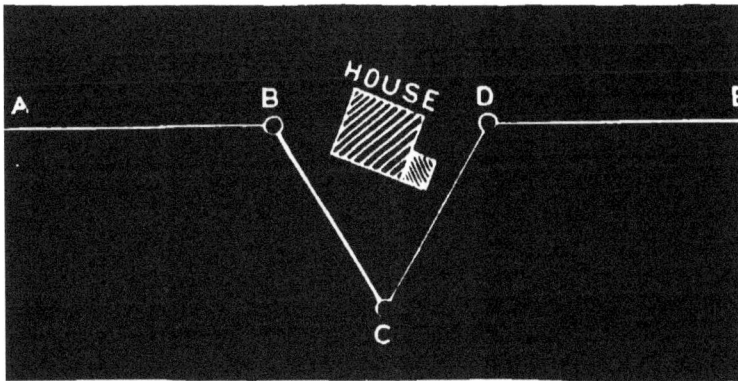

Fig. 2.

1st. *When a tree or house is obstructing the line.* Suppose (in figure 2) A B is the line of survey. At B set up the transit, and taking a back sight on A with the vernier set at 0°, reverse the telescope and turn aside from the line at an angle of 60°, and measure any convenient distance B C. Move to C and turn 60° in the contrary direction, and measure to D the same distance as B C. Then move to D, and turn 60° from C D, prolonged, and D E will be the "line of survey" continued.

2d. *When one end of the line is inac-*

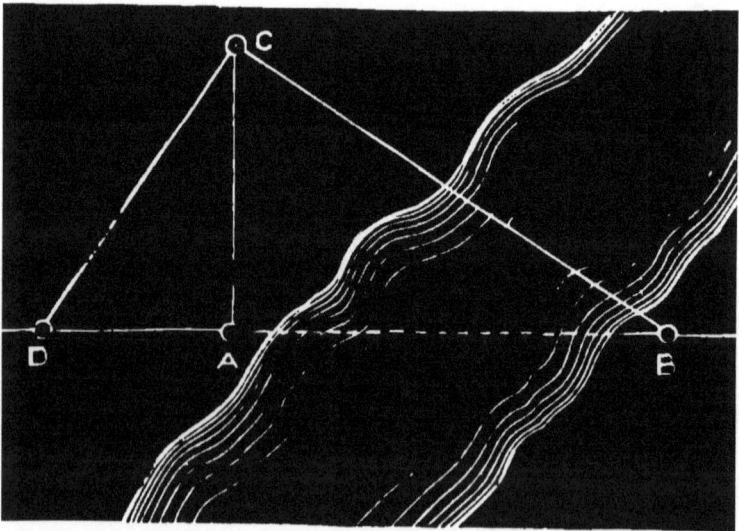

Fig. 3.

cessible. Suppose the line of survey crosses a river (as in figure 3), D A is

the line of survey, and B is inaccessible.
At the point A set off A C perpendicular
to A B of any convenient length. At C
set off a perpendicular to C B, and continue it to a point D in the line of A and
B. Measure D A. Then is $A B = \dfrac{A C^2}{A D}$

These are the most common triangulations
in ordinary practice, but other cases may
occur which will require more complicated
figures. A study of the subject is then
advisable. We have now the line of survey, and want to know the elevations or
levels of the country traversed by that
line. The Leveller and his Rodman have
been following the transit party, as follows: First, setting up the level instrument at some convenient distance about
midway between station 0 and station 8,
he directs the Rodman to hold his rod on
some fixed point, say in this instance, on
top of the curbstone, at the corner of
Madison and Elm Streets, and then sighting through the telescope, he reads how
much his line of sight is above the bottom

of the rod, thus getting the height of the instrument above the curb. Then assuming a datum line, say of 100.00, he adds this reading, which is, say 8.5 feet, to the datum, and calls the height of instrument 108.50 feet. He now proceeds to take a reading at each station, 0, 1, 2, 3, 4, etc., up to station 8, and notes down each reading in his book, as follows:—

LEVEL BOOK.

Station	Rod.	Instr't	Elev.	Remarks.
B. M.	8.50	108.50	100.0	On curb cor. Madison
0	13.3		95.2	and Elm Streets.
1	6.2		102.3	
2	8.9		99.6	
3	9.7		98.8	
4	5.3		103.2	
5	4.2		104.3	
6	2.6		105.9	
7	7.1		101.4	
+50	15.00		93.5	Surface water, Pan-
8	3.8		104.7	ther Creek.

and subtracting these readings from the height of instrument gives the elevation

of each point or station. When it is necessary to move the level, the Rodman
drives a peg into the ground, and gives
a rod on it which the Leveller sights to,
and after reading it very carefully moves
the instrument ahead, sets it up again,
takes a sight at the rod again, and *adds*
this reading to his last elevation (*i. e.* the
elevation of the turning peg); this gives
him a new height of instrument, and he
proceeds as before.

The preliminary line finished, the next
proceeding is to locate it. Curves must
be put in at every angle, or often the
whole direction of the line changed.
Without stopping to explain the com-
pound curve, reversed curve, or any of
the intricacies of location, it is thought
advisable simply to give a rule for in-
serting a curve at any of the angles of
intersection. Suppose it is required to
find the point A or D at which to commence a curve of a given radius.

RULE.—Subtract half the angle A .B D

from 90°, the remainder will be the angle
B C A or B C D. From a table of natural
tangents, take the tangent of B C A, and
multiply it by the given radius, the product will be B A or B D. Now having
calculated the apex distance, and at what
point the curve is to commence, we measure back from the point of intersection

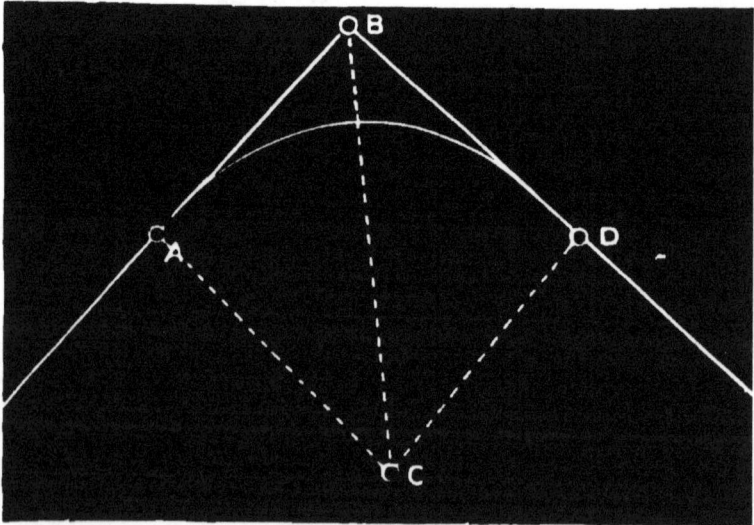

Fig. 4.

that distance, and establish the point of
curve (P C) on the ground. Then to locate
it, the transit is set up at the point of

curve, and a deflection is made for every 100 feet, equal to $\frac{1}{2}$ of the degree of curve. That is, suppose it is required to locate a 6° curve, we first deflect 3°, then for the next station of 100 feet 3° more, and so proceed to the end of the curve, then in

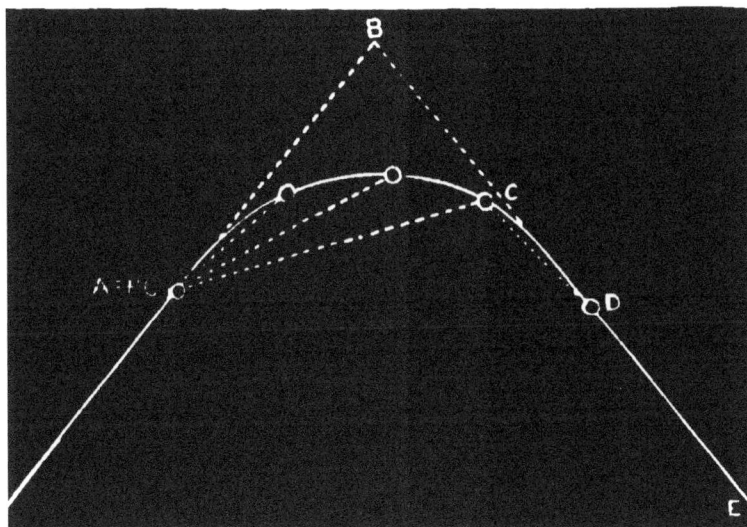

Fig. 5.

order to pass from the end of the curve D on to the tangent D E, place the instrument at D, and sighting back to C, lay off the tangential angle C D B, then B D continued towards E will be the required tangent.

RAILWAY CURVES.

Degree.	Radius.	Vers. sine.
1 deg.	5,730 feet.	1–8 inch.
2 "	2,865 "	3–16 "
3 "	1,910 "	5–16 "
4 "	1,432 "	7–16 "
5 "	1,146 "	1–2 "
6 "	955 "	5–8 "
7 "	819 "	3–4 "
8 "	716 "	13–16 "
9 "	637 "	15–16 "
10 "	573 "	1 1–16 "
11 "	521 "	1 1–8 "
12 "	478 "	1 1–4 "
13 "	441 "	1 3–8 "
14 "	410 "	1 7–16 "
15 "	383 "	1 9–16 "
16 "	359 "	1 11–16 "
17 "	338 "	1 3–4 "
18 "	319 "	1 7–8 "
19 "	302 "	2 "
20 "	287 "	2 1–16 "

The radius of a 1 degree curve equals about $1\frac{1}{12}$ miles. That of any other degree is found by dividing $1\frac{1}{12}$ miles or 5730 feet, by the number of degrees.

EXAMPLE.—Radius 6° curve = 5730 ÷ 6 = 955 feet.

Curvature of the earth is equal to 8 inches per statute mile.

In the foregoing Table of Railway Curves the column headed " vers. sine" corresponds to the middle ordinates of a hundred-foot chord.

Thus in Fig. 5a the chord A B represents one chain (100 feet) and the middle ordinate is the distance, C D. With this distance given in the table it is easy to measure it on the ground, by stretching the chain from A to B, or by setting a stake with the transit at D and 50 feet from A. The distance measured to C will fix the proper point on the curve half way from A to B.

With this brief introduction regarding field operations, a subject has been considered, which in itself would require a volume to be complete. The writer presents only the simplest operations of any Engineer's experience in the field and " on survey," sufficient, however, for the wants of the unscientific man, for whom this work is chiefly intended.

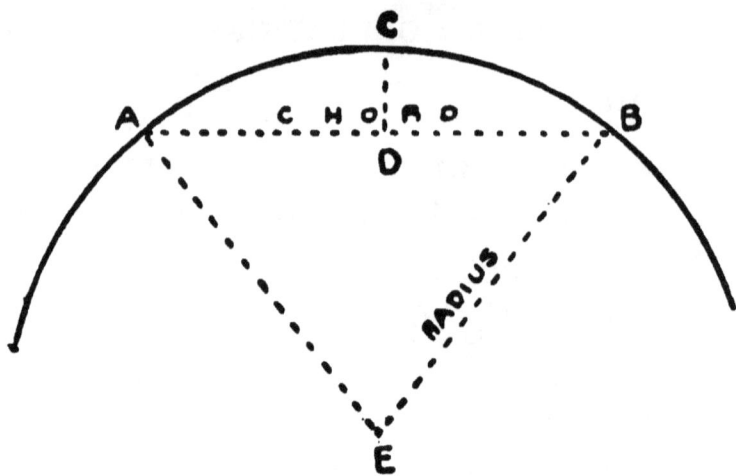

Fig. 5a.

CHAPTER II.

THE first object for consideration in locating a line for a railway is to ascertain, as far as practicable, the probable amount of traffic to be provided for, together with the nature of the same, and the direction in which it will probably run; also what rate of speed is proposed for the rolling stock : if high, the superstructure must be made stronger and heavier, grades must be made easier, and curves lighter, than if a low rate of speed is contemplated. With this information in view, and carefully considered, the Engineer can proceed to locate the line with more intelligence, and a greater degree of success than usually attends the average "locations" of the present day. The first, or PRELIMINARY SURVEY, is made with the view of examining the country through which the contemplated railroad is to pass, and by trial lines, and notes collected from actual survey, to arrive at some probable route, and an approximate estimate of its cost. The points from which the railroad is to start and terminate, are given, and the Engineer's duty consists in LOCATING THE LINE which is to connect them, and to prepare a plan, profile, and estimate of the cost of building a railroad on that line.

This duty involves a great amount of labor, trouble, and annoyance. A determined, resolute Engineer will make his location, entirely relying on his own judgment and skill, making such changes only as he himself thinks right, ignoring completely scheming directors and others having for their only object the filling of their pockets at the expense of the enterprise. It is too often the case that the locating Engineer is besieged by the men along the line, very often directors of the company, and obliged to subject his own better judgment as to what route would be best, to the wishes of an interested party, who is too powerful with the company for the Engineer to differ from. Again a wealthy individual along the line proposes to subscribe liberally to the stock " if the line goes here or there," or any place except the right place, where the Engineer has fixed it. And so very often, owing to such influences, the road is improperly located; large expenses are incurred to correct it; profits melt away on the poorly located line, and the road is not only a failure, but the Engineer's ability is questioned much to his injury. It is surprising how necessary a thing it is for a man to " serve his time" to every profession except Engineering. The moment a railroad is projected, every man 'along the line becomes a good locating Engineer, and viewing with disgust the line as proposed by the Engineer who has devoted a dozen or more years of his life in the practice of his profession, points out what route *he* would take. Still, as before suggested, an Engineer should use his own judgment, ignoring all counter suggestions by unskilful men, acting in perfect fairness to all parties, and with strict fidelity to the company employing him. Much more could be said regarding location, but the writer is necessarily restricted to a simple hand-book, and the temptation to write further cannot be indulged in. Several very valuable works are in print, treating the subject in every detail, and these should be studied carefully before the young engineer undertakes the location of a railway.

If the country through which the line is intended to run is well adapted to the construction, a preliminary survey can be arrived at, and an estimate made in a comparatively short time, regulated, of course, by the length of the road and the difficulties to be encountered. A fair estimate of the cost of a preliminary survey and estimate, is $35.00 per mile, including everything (although much higher charges are made by Engineers of a speculative turn of mind), plan, profile, and estimate. When this has been made and the feasibility of building the road is decided, negotiations should be commenced for the "RIGHT OF WAY," which will determine, in a great measure, whether the road can be built. When it is known that the road is really to be built, and through certain properties, the value of those "estates" immediately assumes gigantic proportions; persons who before the survey was made were willing to give their lands away wholesale to the railroad company, now become strangely reluctant even to sell the smallest morsel. The frailty of human nature is sadly exemplified in these property owners, and a barrier, sometimes insurmountable, is placed in the company's way. In this dilemma, there are two alternatives: one is to revise the line, take another route, and avoid altogether the exorbitant charges; and the other is, the law and courts; the latter are very often appealed to, but only as a last resort, it being exceedingly difficult to find twelve jurymen who are not more or less in close sympathy with the individual, and against the company. An amicable adjustment should be made, if possible, between the parties by an uninterested arbitrator taken from a different section of country. This will be found the easiest and the simplest form of adjustment. Should the property owners be willing to concede sufficient ground for the right of way, a conditional or preliminary agreement is entered into, in the following form :—

Grant of Right of Way and Release of Damages.

—— ——,

To the —— —— *Railroad Co.*

Know all men by these presents, That whereas "THE —— —— RAILROAD COMPANY," a corporation formed under

and in pursuance of the laws of the
State of ——, for the purpose of locating and constructing a Railroad from
—— to —— in said State, have located
or are about to locate their Railroad
through, over, or upon the lands, premises, and property of the undersigned
—— in —— County in the said State, and
for the said purpose are desirous of obtain-
ing the Right of Way in, through, and
upon the said lands and premises.   Now,
therefore, the said —— for and in consideration of the location of the said Railroad
through and upon his lands, and of the
advantages which may accrue to him
therefrom, and also of the sum of one
dollar in hand paid, the receipt whereof is
hereby acknowledged, —— doth hereby,
for himself, his heirs, executors or administrators, give, grant, sell and convey
unto the said " THE —— —— RAILROAD
COMPANY," their successors and assigns,
for the uses and purposes of their Railroad and the construction of works con-

nected therewith, the absolute Right of
Way through, over and upon his said
lands the whole distance of the said Railroad through and over the same, with
the unrestricted right and privilege to
enter upon, locate and construct their
railroad on, over and through his lands
as aforesaid, to such extent as may be
necessary for the location, construction,
opening and use of said Railroad, not
exceeding —— feet in width on and of
the said lands, with such additional width,
however, as may be required at deep
cuttings and embankments, one-half thereof on each side of the *Centre Line* of
the main track of said Railroad as laid
down and established by the said Company on their located route, and the full
liberty to make, maintain and use the
said Railroad over, through and upon the
said lands, with the usual Road-bed,
Slopes, Berms, Ditches, Spoil-banks and
Borrow-pits; and also the right to take
and use any water from springs or streams

upon the said lands, and to conduct and carry water by pipe or otherwise over, through or under the same, and to establish Water-stations thereon.

And the said —— further covenants and agrees with the said Railroad Company, their successors and assigns, that on the said Railroad from —— City to —— City being completed and placed in running order, he, his heirs, executors or administrators, at the proper cost and request of the said Railroad Company, will grant and convey the lands and premises hereinbefore described, and the rights and privileges appurtenant thereto, to the said Railroad Company, their successors and assigns, so long as the same shall be required for the uses of the said Railroad by the said Company, its successors and assigns, —— and in said Instrument of Conveyance to discharge and forever release the said " THE —— —— RAILROAD COMPANY," their successors and assigns, from any further payments for, or on account of the use and occupancy of the said lands and premises, as well as for any and all damages which have accrued or which may hereafter accrue by reason of the location, construction, operating and using of the said Railroad through, over, and upon the lands aforesaid. Provided, however, that the construction of the said Railroad shall be begun within twelve months, and the said Road shall be in operation within two years from this date.

In witness whereof, the said —— hath hereunto set his hand and affixed his seal the —— day of —— A. D. 1877.

—— ——. [SEAL.]

WITNESS:

—— ——.

The following table shows how much ground is required, per mile and hundred feet, for different widths for *Right of Way* purposes.

Table of acres required per mile, and per 100 feet for different widths (Traut-wine).

Width in feet.	Acres per mile.	Acres per 100 feet.	Width in feet.	Acres per mile.	Amount per 100 feet.
20	2.42	.046	31	3.76	.071
21	2.55	.048	32	3.88	.073
22	2.67	.051	33	4.00	.076
23	2.79	.053	34	4.12	.078
24	2.91	.055	35	4.24	.080
" $\frac{3}{4}$	3.	.057	36	4.36	.083
25	3.03	.057	37	4.48	.085
26	3.15	.060	38	4.61	.087
27	3.27	.062	39	4.73	.090
28	3.39	.064	40	4.85	.092
29	3.52	.067	41	4.97	.094
30	3.64	.069	" $\frac{1}{4}$	5.	.094

The value of the ground, of course, will vary in different localities.

After the location of the railroad has been determined upon, the estimates made, and the quantities computed, by calculation, the work is ready for " letting," and the contract for building the road can be made. The following form of contract is taken from a contract made by the Penn. sylvania Railroad.

Form of Contract and Proposal.

—— —— Railroad.

CONTRACT.

Articles of Agreement made and concluded this —— day of —— in the year of our Lord one thousand eight hundred and —— by and between —— of the first part, and the —— railroad company, of the second part, *witnesseth*, that for and in consideration of the payments and covenants hereinafter mentioned, to be made and performed by the said Railroad Company, the said party of the first part doth hereby covenant and agree to construct and finish, in the most substantial and workman-like manner, to the satisfaction and acceptance of the Engineer of said Company, all the graduation, masonry, and such other work as may be required on Section ——, numbered —— of said road: the said work to be finished

as described in the following SPECIFICA-
TIONS, and agreeably to the directions,
from time to time, of the said Engineer
or his assistants, on or before the —— day
of —— in the year one thousand eight
hundred and ——.

SPECIFICATIONS.

1. *Graduation.*—Under this head will
be included all excavations and embankments required for the formation of the
road-bed; cutting all ditches or drains
about or contiguous to the road, the
foundation of culverts, and bridges or
walls; the excavations and embankments
necessary for reconstructing turnpikes or
common roads, in cases where they are
destroyed or interfered with in the formation of the road; and all other excavations or embankments connected with or
incident to the construction of said railroad.

2. All cuttings shall be measured in
the excavations, and estimated by the

cubic yard, under the following heads, viz.: *earth, loose rock, solid rock, tunnel excavation, embankment.*

Earth will include clay, sand, loam, gravel, and all other earthy matter, or earth containing loose stone or boulders, intermixed, which do not exceed in size three cubic feet.

Loose Rock shall include all stone and detached rock lying in separate and contiguous masses, containing not over one cubic yard; also all slate or other rock that can be quarried without blasting, although blasting may be occasionally resorted to.

Solid Rock includes all rock occurring in masses, exceeding one cubic yard, which cannot be removed without blasting.

Tunnel Excavation includes all excavation necessarily taken from the area required to be tunnelled.

3. The road will be graded for a single track, except where otherwise directed

by the Engineer; with side slopes of such inclinations as the Engineer shall, in each case, designate, and in conformity to such breadths, depths, and slopes of cuttings and fillings as may have been, or may hereafter be, determined upon by said Engineer.

4. Earth, gravel, and other materials taken from excavation (except when otherwise directed by the Engineer) shall be deposited in the adjacent embankments, the cost of removing which will be included in the price paid for excavation. It will be understood, therefore, that the excavation price is designed to pay for the excavation, loading, hauling, and dumping in embankments, all material necessarily procured from within the line of the railroad.

Embankment includes all material placed in the embankments of the roadway of the railroad, or common roads which may be crossed or changed in their locations; this material, when taken from the excavations of the railway, will be paid for as embankment when hauled to a distance of 1000 feet; in that case it will be paid for as excavation, and also as embankment; when hauled to a distance less than 1000 feet, it will be paid for as excavation only. In procuring materials for embankment from without the line of the road, the place will be designated by the Engineer in charge of the work: and in excavating and removing it, care *must* be taken to injure or disfigure the land as little as possible. The embankments will be formed in layers of such depth (generally one foot), and the materials disposed and distributed in such a manner as the Engineer may direct, the required allowance for settling being added.

Material necessarily wasted from the cuttings shall be used in widening the banks, or be deposited in the vicinity of the road, according to the directions of the Engineer.

5. The ground to be occupied by the excavations and embankments, together with a space of twelve feet beyond the slope-stakes on each side, or ten feet beyond the berme ditch, where one is required, will be cleared of all trees, brush, and other perishable matter. Where the filling does not exceed two and one-half feet, the trees, stumps, and saplings must be grubbed, but under all other portions of the embankment, it will be sufficient that they be cut close to the earth; no separate allowance will be made for grubbing and clearing, but its cost will be included in the price for excavation.

6. Contractors, when desired by the Engineer in charge of the work, will deposit on the side of the road, or at such convenient points as may be designated, any stone or rock that they may excavate; and if, in so doing, they should deposit material required for embankment, the additional cost, if any, of procuring other materials from without the

road, will be allowed. All stone or rock excavated and deposited as above, together with all timber removed from the line of the road, will be considered the property of the railroad company, and the contractors upon the respective sections will be responsible for its safe keeping until removed by said Company, or until his work is finished.

7. The line of the road, or the gradients, may be changed if the Engineer shall consider such changes necessary or expedient; and for any considerable alterations, the injury or advantage to the contractor will be estimated, and such allowance or deduction made in the prices as the Engineer may deem just and equitable; but no claim for an increase in prices of excavation or embankment on the part of the contractor will be allowed or considered unless made in writing before the work on that part of the section, where the alteration has been made, shall have commenced. The Engineer

may, also, on the conditions last named, increase or diminish the length of any section for the purpose of more equalizing or balancing the excavations and embankments.

8. Whenever the route of the railroad is traversed by public or private roads, commodious passing places must be kept open and in safe condition for use; and in passing through farms the contractor must always keep up such necessary fences as will be needed for the preservation of the crops.

MASONRY.

All masonry will be estimated and paid for by the cubic yard of twenty-seven feet (cubic), and will be included under the following heads, viz..: *culvert masonry, bridge masonry, vertical and slope wall masonry.*

1. *Culvert Masonry.*—All rectangular culverts will be built dry, with a water way of not less than two and a half by

three feet; the abutments will rest on a pavement of stone, set edgewise, of at least ten inches in depth, confined and secured at the ends by deep curb-stones, which must be protected from undermining by broken stone placed in such quantities and position as the Engineer may direct. The abutment walls will not be *less* than two feet thick, and built of good-sized and well-shaped stone, properly laid and bound together by stones, occasionally extending entirely through the walls. The upper course to have at least one-half of the stone headers; and the structures in no case to be less than twelve inches wide; no stone in this course to be less than six inches thick. The covering to be of sound, strong stone, at least twelve inches thick, and to lap its whole width not *less* than ten inches on each abutment. The thickness of the covering, stone, and dimensions of the whole walls to be increased at the discretion of the Engineer.

2. *Bridge Masonry.*—When rock foundation cannot be had for abutments and piers, the masonry shall be started upon hewn timber, sunk to such a depth as to protect it from decay, and to prevent the possibility of underwashing. The timber platforms will be composed of one or more courses, according to the depth of the water, the height of the masonry, or other circumstances of which the Engineer shall judge and determine. The masonry will be of two qualities, either to be adopted at the *discretion* of the Engineer. FIRST QUALITY shall be rock range work. The stone to be accurately squared, jointed, and bedded, and laid in courses of not less than twelve inches thick, nor exceeding twenty inches in thickness, regularly decreasing from bottom to top of pier or abutment. The stretchers shall, in no case, have less than sixteen inches bed for a twelve inch course, and for all courses above sixteen inches, at least as much bed as face; they shall generally

5

be at least four feet in length. The headers will be of similar size as stretchers, and shall hold the size in the heart of the wall that they show on the face, and be so arranged as to occupy one-fifth of the face of the wall, and they will be similarly disposed in the back. When the thickness of the wall will admit of their interlocking they will be disposed in that manner. When the wall is too thick to admit of that arrangement, stones not less than four feet in length will be placed transversely in the heart of the wall to connect the two opposite sides of it. The stone for the heart of the wall will be of the same thickness as those in the face and back, and must be well fitted to their places; any remaining interstices will be filled with ordinary masonry. The face stones will, with the exception of the draught, be generally left with the face as they come from the quarry unless the projections above the draught should exceed two inches, in which case they must be roughly scabbed down to that point. The abutments or piers, and such portions of them as the Engineer may direct, shall be covered with a course of coping not less than ten inches thick, well-dressed, and fastened together with clamps of iron.

The SECOND QUALITY of bridge masonry will be rubble work, laid in irregular courses, and will consist of stone containing generally six cubic feet each, so disposed as to make a firm and compact work; and no stone in the work shall contain less than two cubic feet, except for filling up the interstices between the large blocks in the heart of the wall; at least one-fifth of the face shall be composed of headers, extending full size four feet into the wall, and from the back the same proportion and of the same dimensions, so arranged that a header in the back shall be between two headers in the face. The corner stones shall be neatly hammer-dressed so as to have horizontal beds and vertical joints.

3. *Vertical and Breast Walls.* — The walls will be good, dry rubble work, the stones to be of such dimensions and laid with such batter as the Engineer may direct.

4. BRICK WORK.—Where bricks are used in piers or abutments of arched or open bridges or tunnels, they shall be made of the best clay, well tempered, moulded, and burnt. The sizes after burning to be nine inches long, four and one-quarter inches wide, and two and one-half inches thick; laid in the best hydraulic mortar, grouted full every three courses, and made with such proportions of cement and sand as the Engineer may direct. The materials for the mortar to be furnished by the contractor. The joints to be of such thickness, and the bond to be of old English or Flemish, or such other character as the Engineer may prescribe, either for the

walls or arches. No bats, cracked, broken, or salmon brick to be used in the work.

The quality of the stone or brick of which the masonry shall be built must be well suited to the kind of structure in which it is used, according to the judgment of the Engineer. All masonry, whether of stone or brick, to be estimated and paid for by the cubic yard of twenty-seven cubic feet. Such portions of the masonry as the Engineer may require to be laid in lime mortar or hydraulic cement, will be so laid.

5. The prices per cubic yard for masonry shall in every case include the furnishing of all materials, excepting lime and cement, the cost of scaffolding, centering, etc., and all the expenses attending the delivery of the materials, and all risks from floods and otherwise.

6. No charge shall be made by the contractor for hindrances or delay from any cause in the progress of any portion of

the work in this contract, but it may entitle him to an extension of time allowed for completing the work, sufficient to compensate for the detention; to be determined by the Chief Engineer, provided he shall give the Engineer in charge immediate notice in writing of the cause of the detention.

Nor shall any claim be allowed for extra work, unless the same shall be done in pursuance of an order from the Engineer in charge, and the claim made at the first settlement after the work was executed, unless the Chief Engineer at his discretion should direct the claim or such part as he may deem just and equitable to be allowed.

And the said —— Railroad Company doth promise and agree to pay to the said party of the first part, for completing this contract, as follows, viz. :—

For earth excavation,   — cents per cub. yard.
For loose rock excavation, — "   "   "
For solid rock excavation, — "   "   "

Tunnel excavation, — cents per cubic yard.
For embankment, — " " "
For bridge masonry,
    1st quality, — " " "
    2d quality, — " " "
For rectangular culverts, — " " "
For brick arches laid in
    cement, — " " "
For paving in foundations, — " " "
For vertical walls or breast
    walls, — " " "
For timbers in foundations,
    in position, — " per cub. foot.
For timber worked in
    trestling, — " per 1000 ft. B. M.
For timber in bridges, — " " "
For workmanship of timber
    in bridges, — " " "
For wrought iron, in
    position, — " per pound.
For cast iron, in position, — " " "
Any and all other items at the Engineer's estimate.

On or after the first day of each month during the progress of this work, an estimate shall be made of the relative value of the work done, to be judged of by the Engineer, and upon his certificate of the amount being presented to the Treasurer of the Railroad Company the amount of said estimate shall be paid to the party of the first part, at such time and place as the said Treasurer may designate; and when all the work embraced in this contract is completed agreeably to the specifications, and in accordance with the directions, and to the satisfaction and acceptance of the Engineer, there shall be a final estimate made of the *quality*, *character*, and *value* of said work, agreeably to the terms of this agreement, when the balance appearing due to the said party of the first part shall be paid to —— upon —— giving a release, under seal of the said Railroad Company, from all claims or demands whatsoever, growing in any manner out of this agreement. It is further covenanted and agreed between the said parties that the said party of the first part shall not let or transfer this contract to any person (excepting for the delivery of materials) without the consent of the Engineer, but will give personal attention to the work. It is further agreed that the work embraced in this contract shall be commenced within —— days from this date and prosecuted with such force as the Engineer shall deem adequate to its completion within the time specified, and if at any time the said party of the first part shall refuse or neglect to prosecute the work with a force sufficient in the opinion of the said Engineer for its completion within the time • specified in this agreement, then, and in that case, the Engineer in charge, or such other agent as the Engineer may designate, may proceed to employ such a number of workingmen, laborers, and overseers as may, in the opinion of the said Engineer, be necessary to insure the completion of the work within the time hereinbefore limited, at such wages as he may find it necessary or expedient to give, pay all persons so employed, and charge over the amount so paid to the party of the first part, as for so much

money paid to said party of the first part
on this contract; or the said Engineer
may, at his discretion, for the failure to
prosecute the work with an adequate
force, for non-compliance with his directions in regard to the manner of constructing it, or for any other omission or
neglect of the requirements of this agreement and specifications on the part of the
party of the first part, declare this contract, or any portion or section embraced
in it, forfeited, which declaration and forfeiture shall exonerate the said ——
Railroad Company from any and all obligations and liabilities arising under this
contract, the same as if this agreement
had never been made, and the reserved
percentage of —— upon any work done
by the party of the first part may be
retained forever by the said Railroad
Company. And it is mutually agreed
and distinctly understood that the decision of the Chief Engineer shall be
final and conclusive in any dispute which

may arise between the parties to this agreement, relative to or touching the same, and each and every of said parties do hereby waive any right of action, suit · or suits, or other remedy in law or otherwise, by virtue of said covenants, so that the decision of said Engineer shall, in the nature of an award, be final and conclusive on the rights and claims of said parties.

IN WITNESS WHEREOF, the President of the —— Railroad Company hath signed the same, and caused the corporate seal of said Company to be attached, and the said —— ha— hereunto set —— hand— and seal— the day and year first above written.

—— ——. [SEAL.]
—— ——. [SEAL.]
—— ——. [SEAL.]
—— ——. [SEAL.]

WITNESS:

—— ——.

RAILROAD PROPOSALS.

	$	cts.
Excavation per cubic yard, earth,		
" " " loose rock,		
" " " solid "		
" " " tunnel,		
Embankment per cubic yard,		
Masonry (per yard of 27 cubic feet),		
" of bridges, 1st quality,		
" of bridges, 2d quality,		
" of arches, stone,		
" of arches, brick,		
" of rectangular culverts,		
" of paving in foundations,		
" of vertical or breast walls,		
Timber (per 1000 feet, B. M.)		
" in bridges,		
" in trestles,		
Workmanship (per 1000 feet, B. M.)		
" in bridges,		
" in trestles,		
Timber per cubic foot in foundations,		
Iron per pound, in position,		
Wrought iron per pound, in position,		
Cast " " " "		

The undersigned hereby propose to the —— Railroad Company to do all the work on either or all of the sections to which prices are affixed in the schedule, according to the conditions and specifications contained in the printed form of contract, a copy of which is annexed; and, on the acceptance of this proposal for all or either of the above sections do —— hereby bind —— to enter into and execute a contract in said form for the prices above named.

—— 18——

Proposer's Residence,

Nearest Post-office. Signed,

—— ——.

The above form of Proposal is filled in by the contractor and attached to the blank form of contract. This constitutes his bid for the work, and is considered, with others, by the railway officials on some previously advertised day.

CHAPTER III.

THE work required to bring the natural formation of the ground to the grade lines of the proposed railroad is called GRADING, and embraces all the cutting and embankment required. Being by far the most expensive part of the enterprise, close attention must be given to it in every estimate that is made. It is divided into two classes; *excavation* and *embankment*. It is very desirable, as far as practicable, for the Engineer to so arrange his grade lines that all the cuttings will yield sufficient material to form all the embankment; the nearest approach to this happy equality of affairs generally shows the cheapest possible line; but there are a great many reasons why such a course is not practicable: for example,

78

the haul may be too long to bring the material from the cutting to the place of embankment, consuming much time and money; in this case "BORROW PITS" are necessary—that is, material must be borrowed from other points than the excavations of the line to form the banks. And in the case of excavation a "WASTE" is organized, or the material from the excavations instead of being used to form the banks is wasted, i. e., dumped into any convenient hollow or ravine. This borrowing and wasting is a very expensive business, and as little of it should be allowed as possible; a contractor should haul at least 1000 feet before any idea of borrowing is allowed. The steepest or MAXIMUM GRADE of a railroad is to be determined by the rolling load that is to pass over it. On the Phila. and Reading R. R., a train of 170 loaded cars, each car carrying 5 tons of coal, is hauled with comparative ease on a level grade at the rate of 12 miles an hour. The following table shows the number of feet per 100 feet, ascending or descending grade, for each degree and minute of the angle of inclination up to 5 feet per 100 feet.

Table of Grades per 100 Feet for Each Degree to 5 Feet.

Deg.	Min.	Per 100'.	Deg.	Min.	Per 100'.	Deg.	Min.	Per 100'.	Deg.	Min.	Per 100'.
0	1	.0291	0	30	.8727	1	0	1.7455	1	56	3.3758
	2	.0582		31	.9018		2	1.8038		58	3.4341
	3	.0873		32	.9309		4	1.8620	2	0	3.4924
	4	.1164		33	.9600		6	1.9202		2	3.5506
	5	.1455		34	.9891		8	1.9784		4	3.6087
	6	.1746		35	1.0182		10	2.0366		6	3.6669
	7	.2037		36	1.0472		11	2.0948		8	3.7250
	8	.2328		37	1.0763		12	2.1530		10	3.7833
	9	.2619		38	1.1054		14	2.2112		12	3.8416
	10	.2909		39	1.1345		16	2.2694		14	3.8999
	11	.3200		40	1.1636		18	2.3277		16	3.9581
	12	.3491		41	1.1927		20	2.3859		18	4.0163
	13	.3782		42	1.2218		22	2.4441		20	4.0746
	14	.4073		43	1.2509		24	2.5023		22	4.1329
	15	.4364		44	1.2800		26	2.5604		24	4.1911
	16	.4655		45	1.3090		28	2.6186		26	4.2494
	17	.4946		46	1.3381		30	2.6768		28	4.3076
	18	.5237		47	1.3672		32	2.7350		30	4.3659
	19	.5528		48	1.3963		34	2.7932		32	4.4242
	20	.5818		49	1.4254		36	2.8514		34	4.4826
	21	.6109		50	1.4545		38	2.9097		36	4.5409
	22	.6400		51	1.4837		40	2.9679		38	4.5993
	23	.6691		52	1.5128		42	3.0262		40	4.6576
	24	.6982		53	1.5419		44	3.0844		42	4.7159
	25	.7273		54	1.5710		46	3.1427		44	4.7742
	26	.7564		55	1.6000		48	3.2010		46	4.8325
	27	.7855		56	1.6291		50	3.2592		48	4.8908
	28	.8146		57	1.6583		52	3.3175		50	4.9492
	29	.8436		58	1.6873		54	3.3758		52	5.0075
				59	1.7164						

To get the grade in feet *per mile*, multiply the figures given in the column headed PER 100' by 52.80. Thus, in the Table, angle 2° 52' = 5.0075 × 52.80 = 264.39 feet per mile.

Where the trade will not be very heavy and the trains light, much steeper grades can be used, saving very materially in the first cost of the road, by making the excavations less heavy, and the banks not so high. On the Cumberland and Pennsylvania R. R. there are grades of 186 feet to the mile, and on some of the recently constructed narrow-gauge railroads, the writer has been informed of grades of 4 feet in the 100 or 211 feet to the mile. Often a great deal of unnecessary expense is incurred in the endeavor to preserve a uniform grade, which could be saved by building a "surface road" or establishing the grades by frequent changes, so as to conform to the natural surface of the ground.

6

The accompanying illustration shows the manner in which the surface of the ground is plotted on profile paper and the grade of the railway established.

A good locomotive weighing 27 tons on the drivers can haul up a grade of

5 feet to the mile,	1150 tons.	
10 " "	939 "	
20 " "	686 "	
30 " "	536 "	
40 " "	437 "	
50 " "	367 "	
60 " "	315 "	
70 " "	275 "	
80 " "	242 "	At a speed
90 " "	216 "	of 8 to 12 miles per
100 " "	194 "	hour.
110 " "	175 "	
120 " "	159 "	
130 " "	146 "	
140 " "	134 "	
150 " "	123 "	
160 " "	113 "	
170 " "	105 "	
180 " "	98 "	

The grade being established to the best possible advantage, the next step is to

Profile of railway showing excavations and embankments.

provide for "STAKING OUT" THE WORK.
This is the work of the "constructing
corps." The places to be excavated are
marked on the ground by driving stakes
every 50 or 25 feet along the entire line,
and the number of feet of cutting or embankment is marked on them with red
chalk or " kehl." The side or slope stakes
are then set, which indicate the position
of the edge of the slope. For the benefit
of young Engineers perusing this work,
who are not familiar with staking out
work, the writer, digressing from the
general plan of this treatise, takes the
occasion to introduce a simple rule for
"staking out," which he has used frequently during his practice in construction work, as follows: When the natural
surface of the ground is level, *Add the
cut in feet and decimals, multiplied by
the slope, to one-half the road-bed.* Thus,
for example, suppose the depth of the cut
required to be 20.3 feet, the slope to be
1½ to 1, and the road-bed 13 feet, then,

by the rule given above, we have (20.3 × 1½) + 6.5 = 36.95 feet, which is the distance from the centre line to the edge of the slope on either side of the line. But suppose that the natural surface of the ground is not level, then assume a point on the ground (apparently right) find its height above grade with the level, multiply this by the slope, and add one-half the road-bed; see how near this calculation comes to the measured distance from the centre to the assumed point; if not right, a second trial will fix the point. If the natural surface of the ground is very much inclined, stake out the upper side only, and allow the lower side to assume its own shape. The section at each station should be plotted on cross-section paper from which the areas are calculated. With this information obtained the cubical contents of each 100 feet of excavation or embankment can be ascertained by the prismoidal formula as given on page 94. The ground, being staked out, is then loosened by picks or ploughs, the latter generally being much cheaper; a single plough with two horses and men will loosen up from 200 to 300 yards of stiff soil per day at about $1\frac{1}{2}$ cents per yard; with the pick, a day's work is about 25 yards, or with labor at $1.00 per day 4 cents per yard. Light soils will average about one-half the above, while pure sand requires very little labor, say $\frac{1}{2}$ cent per yard. After loosening the earth it must be shovelled aside, or into carts or wheelbarrows, and then moved away. A cart will hold about $\frac{1}{3}$ of a cubic yard, measured in place. A man can shovel and load a cart in five minutes, or for a day's work of 10 hours, 120 loads, or 40 cubic yards of light material, but some deduction must be made for delays, etc., which would place the average at about $\frac{1}{2}$ or 20 cubic yards of material one man can load into a cart per day. Assuming the labor at $1.00 per day, the cost of shovelling into carts

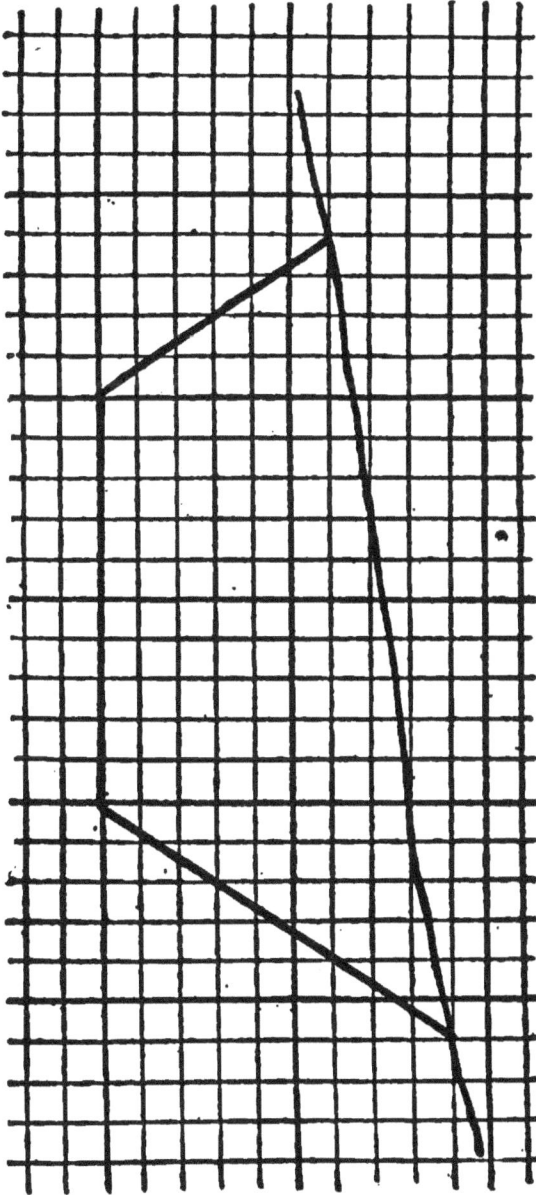

Cross section of railway cutting.

will be about 5 cents a yard. A cart itself weighs about ½ a ton. After the material is loaded into carts it must be hauled away and dumped where it is needed in forming bank, then the cart must return and be loaded again, all of which takes time. Trautwine says: "The average speed of horses in hauling is about 2⅓ miles per hour, or 200 feet per minute, which is equal to 100 feet of trip each way, or to 100 feet of *lead*, as the distance to which *the earth is hauled* is technically called. Besides this, there is a loss of four minutes in every trip, whether long or short, in waiting to load, dumping, turning, etc. Hence every trip will occupy as many minutes as there are lengths of 100 feet each in the lead, and four minutes besides; therefore, to find the number of trips per day over any average lead, we divide the number of minutes in a working day by the sum of 4 added to the number of 100-feet lengths contained in the distance to which the earth has to be removed, that is—

$$\frac{\text{The number (600) of minutes in a day}}{4 + \text{the number of 100 feet-lengths in the lead,}} = \begin{array}{l}\text{The} \\ \text{number} \\ \text{of trips} \\ \text{or loads} \\ \text{per day,} \\ \text{per cart.}\end{array}$$

And since $\frac{1}{3}$ of a cubic yard, measured before being loosened, makes an average cart-load, the number of loads divided by 3 will give the number of cubic yards removed per day by each cart, and the cubic yards divided into the total expense of a cart per day will give the cost per cubic yard for hauling. In leads of ordinary length, one driver can attend to 4 carts, which at $1.00 per day is 25 cents per cart. When labor is $1.00 per day, the expense of a horse is about 75 cents, and that of the cart, including harness, tar, repairs, etc., 25 cents, making the total daily cost per cart $1.00. The expense of the horse is the same on Sundays and on rainy days as when at work, and this consideration is included in the 75 cents. Some contractors employ a greater number of drivers, who also help to load the carts, so that the expense is about the same in either case.

Example.—How many cubic yards of loam, measured in the cut, can be hauled by a horse and cart in a day of 10 working hours (600 minutes), the lead or length of haul of earth being 1000 feet (or 10 lengths of 100 feet); and what will be the expense to the contractor for hauling per cubic yard, assuming the total cost of cart, horse, and driver, at $1.25 ?

Here

$$\frac{600 \text{ minutes}}{4 + 10 \text{ lengths of 100 feet}} = \frac{600}{14} = 43 \text{ loads.}$$

And

$$\frac{43 \text{ loads}}{3} = 14.3 \text{ cubic yards.}$$

And

$$\frac{125 \text{ cents}}{14.3 \text{ cubic yards}} = 8.74 \text{ cents per cubic yard.}"$$

After the material is hauled away and dumped into position, it is necessary to have it nicely spread in layers on the bank and levelled off. Still quoting from Trautwine: "A bankman will spread from 50 to 100 yards of either common loam or any of the heavier soils, clays, etc., depending on their dryness. This, at $1.00 per day, is 1 to 2 cents per cubic yard; and we may assume 1½ cents as a fair average for such soils, while 1 cent will suffice for light sandy soils." Add to the above items, say 2 cents per cubic yard for keeping the haul in order, and 5 cents per yard for contingencies, and we have the AVERAGE COST OF EXCAVATING ONE CUBIC YARD OF EARTH, and placing it in position in the bank, as follows:—

Loosening by pick,	4.00 cts. per cu. yd.	
Loading into carts,	5.00 "	"
Hauling 1000 feet,	8.74 "	"
Spreading into layers,	1.50 "	"
Keeping *haul* in order,	2.00 "	"
Various contingencies,	5.00 "	"
Total cost to contractor,	26.24 "	"
Add contractor's profit, 10%,	2.62 "	"
Total cost to company,	28.86 "	"

If the material is hauled away by men and wheelbarrows, the cost will exceed the foregoing by about 30 per cent.

"A CUBIC YARD OF ROCK IN PLACE, before being blasted, will weigh about 1.8 tons, if sandstone or conglomerate (150 pounds per cubic foot), or 2 tons if good compact granite, gneiss, limestone, or marble (168 pounds per cubic foot)." With labor at $1.00 per day, a fair estimate for LOOSENING SOLID ROCK would be about 50 cents per cubic yard, and for loose rock, say 30 cents per cubic yard, while the cost of loading and hauling away would be about 25 cents per cubic yard, or the total cost to the contractor for loosening, loading, hauling away, and dumping, say with a haul of 1000 feet, would be about 75 cents per cubic yard measured in place. Add to this 10 per cent. for contractor's profit, and we have the ACTUAL COST TO THE COMPANY, at 82.5 cents per cubic yard of solid rock. Loose rock will cost about 55 cents to the contractor, or 60.5 cents to the company. In a mile of single track railroad through a rolling country it is safe to estimate the earth excavation at about 15.000 cubic yards, which, at say 30 cents per yard, would amount to $4500. Also estimate, for about 1500 cubic yards of ᛫ solid rock, at say 85 cents per cubic yard, which would amount to $1275. And 1500 cubic yards of loose rock, at say 60 cents per yard, amounting to $900, which gives us a total for the cost of earthwork PER MILE of single track railroad of $6675. Circumstances, of course, will vary these figures considerably—they are based on average figures, taken from several estimates of roads already built, through a rolling country, but one presenting no engineering difficulties; the maximum haul is considered 1000 feet in length. If the cutting is over 10 or 12 feet in depth, the present day contractor usually uses a steam shovel. This machine is capable of scooping up from 1

to 2 cubic yards of earth at each movement, or about 3 to 6 yards per minute.
One of these machines costing from
$6000 to $8000 will excavate from 500
to 1500 cubic yards of loosened earth
or gravel daily. The cost of operating
one of these shovels is placed at $30 per
day, or about one-half the cost per cubic
yard of hand shoveling when reduced
to a basis of $1.00 per day for labor.
The rule for calculating the cubic contents of excavation is contained in the
following PRISMOIDAL FORMULA. "*Add
together the areas of the two parallel ends
of the prismoid, and four times the area of
a section half way and parallel to them;
and multiply the sum by one-sixth of the
length of the prismoid, measured perpen-
dicularly to its two parallel ends.*" In
railroads, the prismoids are generally 100
feet long, it is therefore easier to multiply
the sum of the areas in square feet by 100,
and divide the product by 6.

Quantity of Earths equal to a Ton.

Sand, river, as filled into carts, 21 cubic feet.
Sand, pit   "   "   22   "
Gravel, coarse,   "   "   23   "
Marl,   "   "   28   "
Clay, stiff,   "   "   29   "
Chalk, in lumps, "   "   29   "
Earth, mould,   "   "   33   "

The whole subject is ably handled by Trautwine, on " Excavation and Embankment." From the same author the following on Tunnel excavation is adapted by the writer. In making tunnels for railroads they should, if possible, be straight, especially when there is but a single track, inasmuch as collisions or other accidents in a tunnel would be particularly disastrous. A tunnel should not be made unless the depth of cutting exceeds 60 feet. Firm rock of moderate hardness, and of a durable nature, is the most favorable material for a tunnel, especially if free from springs and lying in horizontal strata. In soft rock, or in shales (even if hard and firm at first), or in earth, a lining of hard brick or masonry in cement is necessary. A tunnel should have a grade or inclination in one direction for ease of future drainage and ventilation. No special arrangement is necessary for ventilation, either during construction or after, if the length does not exceed 1000 feet; but beyond that, generally during construction either shafts are made, or air is forced into the tunnel through pipes from its ends. But after the work is finished nothing of the kind generally is necessary.

SHAFTS generally cost from one and a half to three times as much per cubic yard as the main tunnel, owing to the greater difficulty of excavating and removing the material, and getting rid of the water, all of which must be done by hoisting. Their sectional areas commonly vary from about 40 to 100 square feet. In excavating the tunnel itself, a HEADING or passage-way 5 or 8 feet high, and 3 to 12 feet wide is driven and maintained a short distance (10 to 100 feet or more), according to the firmness of the material, in advance of the main work. In rock the heading is just below the top of the tunnel, so that the men can conveniently drill holes in its floor for blasting; but in earth, the heading is driven along the bottom of the tunnel, that being the most convenient for enlarging the aperture to the full tunnel size by undermining the earth and letting it fall. In earth, the top and sides of the heading, as well as the tunnel, must be carefully prevented from caving in before the lining is built, and this is done by means of rows of vertical rough timber, props, and horizontal caps or overhead pieces, between which and the earth rough boards are placed to form temporary supporting sides and ceiling to the excavation. The props and caps are placed first, and the boards are then driven in between them and the earthen sides of the excavation. These are gradually removed as the lining is carried forward. THE LINING, when of brick, is usually from 2 to 3 bricks thick (17 to 26 inches) at bottom, and from 1½ to 2½ bricks thick at the top, and when of rough rubble in cement, about half again as thick. It is important that the bricks or stone should be of excellent hard quality, and laid in good cement. The bricks should be moulded to the shape of the arch. As the lining is finished in short lengths, and before the centres are removed, any cavities or voids between it and the earth should be carefully and compactly filled up. Even in rock if much fissured, or if not of durable character, as common slate, lining is necessary.

THE CROSS SECTION of a single-track railroad tunnel in the clear of everything, and for cars of 11 feet extreme width, should not be less than about 15 feet wide by 18 feet high; nor a double track one less than 27 feet wide by 24 feet high, unless in the last case the material is firm rock, in which a high arch is not necessary for lining. The roof may then be much flatter, so that a height of 20 feet will answer. With cars of 10 feet extreme width, the width of the tunnel may be reduced to 25 feet; or with cars 9 feet wide, to 23 feet. The rate of DAILY PROGRESS from *each* face of a tunnel varies from 18 inches to 9 feet of length per 24 hours with three relays of workingmen. From $1\frac{1}{2}$ to 3 feet may be taken as an average. If the tunnel is through earth the construction of the lining about makes up for the slower excavation of one in rock. In rock, with labor at $1.00 per day, the cost will usually vary with the nature of the rock from $2.00 to $5.00 per cubic yard for the main tunnel; and from $3.00 to $10.00 for the heading; while shafts will average about 50 per cent more than heading. As the sides and roof of a tunnel are very *roughly* blasted, the contractor takes out more material than would be given him if measured in *the clear*. Allowance should be made him for this, or the mode of measurement clearly stated in the specifications. Before commencing a tunnel trial shafts should be sunk to ascertain the nature of the material. In long ones the greatest care and accuracy are necessary for preserving the line of direction, so that the work from both ends shall meet properly in the centre. The cost of a single track tunnel will range from $30 to $75 per foot of length.

On the Reading R. R., during its construction, three tunnels were cut at the following cost per *lineal foot:*—

Black Rock Tunnel, near Manayunk, $90.
Flat   "   "   "   Phœnixville, 130.
Port Clinton   "   82.

which gives an average cost of $101 per lineal foot. The first two named are double track tunnels, and the latter a single track only. The Flat Rock tunnel is 1932 feet long, worked with 6 shafts, and artificial means were used for

Railroad tunnel.

ventilation. During the construction of the Black Rock tunnel, the work progressed at the rate of 100 feet in 52 days, while the Port Clinton tunnel was excavated at the rate of 100 lineal feet in 48 days. It is well to avoid a tunnel if it can possibly be done: a number of trial lines should be run, and shafts should be sunk, before even deciding upon building one. Several tunnels formerly built at great expense have been abandoned within recent years. In one case it was found preferable to build two bridges and revise several miles of operating railroad, and in another instance to maintain an open cut of 30 to 40 feet in depth than to continue using the tunnels.

CHAPTER IV.

PERMANENT WAY.

AFTER the excavations and embankments have been completed, and before the track is laid, the BALLAST should be put on the grading. It consists of from 8 to 28 inches of loose, hard material, of sand, gravel, or good hard broken stone —furnace slag has been used very successfully on some American railroads— reduced to such a size that any piece will go through a ring of two inches diameter. The quantity required will depend on the nature of the material which forms the cuts and banks, and may vary in depth from 1 to 3 feet. Ballasting is necessary to drain the road-bed properly. If the sills or cross-ties should be laid directly on the grading without any intermediate ballast, which unfortunately is often done in the hurry to get the road in operation, —they are laid on soft clay, which in wet weather washes away from underneath them; or else on solid rock bottom, say of excavations, of that nature which presents such a rigid base as to destroy the rails. The ballast on a railroad gives the track a certain amount of elasticity, absolutely necessary to carry the train, which passes over it, with any degre of safety. Frequently the cross-ties and iron are laid on the sub-grade temporarily, only laying the iron so that the ballast can be hauled to the place required by construction trains, and so save the expense of carting the material. Such cases are admissible, but when the first construction train passes over the rails safely, the temptation to let a freight, and then a passenger train, over the line also, is very seldom resisted, and much injury results in the form of badly bent iron, occasionally an upset engine, and sometimes loss of human life. Too little attention is paid to ballasting, excepting upon some of our leading railroads, which make it a prominent feature of construction.

In nearly every instance in which the rails, frogs, and switches show unusual signs of wear, the cause can be traced to dirt, or often, to no ballast. The sand ballast of the South makes a good elastic road-bed. Some of the Southern railways have no other ballast except pure clean sand, and apparently it is sufficient. On the Pennsylvania R. R. the ballast is broken stone, sloping from the sills to the sub-grade, and such attention given to it that care is even taken to have the edges of the ballast laid with a line! The following TABLE gives the number of cubic yards of ballasting required for *one mile* of single-track railroad. Slopes of the ballast 1 to 1, the depth from 12 to 30 inches, and the top width—that is, the width of the road-bed—from 10 to 12 feet.

Table of Ballasting.

Depth in inches.	Top width in feet.			
	10.	11.	12.	
12	2152	2347	2543	Cubic yards.
18	3374	3667	3960	" "
24	4694	5085	5474	" "
30	6111	6600	7087	" "
				" "

The cost of breaking stone for ballast, exclusive of the cost of the stone, if done by hand, will be about $700 per mile of single track railroad. The stone can generally be procured from the excavations along the line of the road; if not, the cost will be about $1500 per mile to buy and put in position. The hard slag or refuse, from blast furnaces, makes an excellent ballast, its extreme brittleness renders it easy to reduce to the proper size, and it does not crumble to dust as softer material will do; the water finds its way through it with ease, and as it never packs closely against the CROSS-TIES, the water will not decay as quickly, the air having free access to the wood.

When railroads were first built in England, the rails were firmly bolted to blocks of granite, which were imbedded in the grading; this gave a durable magnificent road-bed, and one over which an engine could pull a greater load than over our timber cross-ties, but every block of stone under the rail had the same effect on it that an anvil would have to a piece of iron on which the smith is using a sledge; the heavy engines when running at any considerable rate of speed battered down the heads of the rails, particularly at the joints, so badly, that the stone had to be removed and timber put in its place to give an elastic road-bed. STRINGERS placed under the rails, running in the same direction, their entire length, were used, but not successfully; the timber would rot in places, and it was found very expensive and inconvenient in repairing to be obliged to handle such large timber.

THE CROSS-TIES as now used on almost all American railroads, excepting a few Southern roads, which are laid with the longitudinal stringers, consist of timbers laid across the ballast at right angles to the line of the road, and usually measure 9 feet long, 7 inches deep, and 8 inches in width, and are usually trees cut down and roughly hewn on the top and bottom sides. A good hard wood is necessary, to prevent the rails from sinking into it. In England, the double-headed rail is used, that is, the top and bottom of the rail are of the same shape, and it has no flat base like the American rail. This necessitates iron chairs placed on each cross-tie, in order to hold the rail in position. In this country the base of the rail rests directly on the cross-ties, and hard wood is necessary to prevent the crushing of its fibres by the rail. White oak, chestnut, locust, and cedar make good cross-ties; elm can be used if of good quality; whatever timber abounds along the line of the road, if at all hard wood, can be used to greater advantage than any timber which has to be brought from a distance. The seasoning and preparing of cross-ties for railroads has received great attention from many eminent Engineers, and many attempts have been made to prepare the cross-ties previous to laying, so as to prevent or arrest the natural decay of the wood. Some years ago the cross-ties used on the Phila. and Reading R. R. were notched at the points where the rails crossed them, and their ends dipped in coal tar; by this process it was supposed it would preserve the ends from decay. Since then "BURNETTIZING" has been tried on the same road, a process by which the ties were thoroughly saturated with a solution of zinc. Neither of these gave the desired result, and both have been abandoned. The cost of Burnettizing a cross-tie was 25 cents, equal to one-half of its original cost. SAWED CROSS-TIES are often used, but only on trestle work or bridges, although the writer knows of cases where sawed ties are laid the entire length of the railroad. The following table gives the NUMBER OF CROSS-TIES required to *one mile* of single track railroad, laid in the following order.

18 inches from centre to centre,			3520 ties.
21 "	"	"	3017 "
24 "	"	"	2640 "
27 "	"	"	2348 "
30 "	"	"	2113 "

A fair ESTIMATE FOR CROSS-TIES is from 40 to 50 cents apiece, delivered on the line of the road, or about $1700 per mile of single track road.

The history of the *rail* is identical with the history of tramways. Wooden rails were used at New Castle, in 1602.

In 1716, flat pieces of iron were nailed to the wooden rails.

In 1776, cast-iron rails, with upright flange, were laid on wooden sleepers.

In 1789, Loughborough's cast-iron edge rail, with flanges on the wagon wheels.

In 1793, stone bearings were substituted for wooden sleepers.

Wrought-iron bars two or three inches in thickness, spiked to longitudinal sleepers, were then used in connection with flanged wheels.

Wyatt's cast edge rail, leaving an oval section, was then used in connection with grooved wheels in 1800.

Jessop used this rail in 1789, and added the chair—a block of iron—slotted to receive the ends of adjacent rails. The wheel had a tread of $2\frac{1}{2}$ inches, and a flange to keep it on the rail; the sleepers were of wood.

1803. Woodhouse's hollow rail, with a channel for the rounded edge of the wheel.

1805. The fish-bellied rail at Penrhyn.

1810. A square-bodied cast rail.

1811. Blenkinsop's rock rail.

1816. Gosh and Stephenson's flanged rail, which was a lapping continuous rail.

1817. Hawk's cast-iron face on a wrought-iron base.

1820. Birkenshaw's, of Bedlington, Durham, wrought-iron face on a cast-iron base; he also invented the rolled rail, the iron, while hot, being passed between grooved rollers of the required pattern. This rail, with many modifications, is now used on the different railroads in this country.

As before stated, some few Southern roads in this country are using the old STRAP RAIL, a flat bar of wrought iron resting in its entire length on wooden stringers, but they are only so used for light logging roads in the lumbering districts, and would not be suitable for use with our modern rolling stock. In the olden times this was the rail in general use in America, and many accidents occurred by reason of the ends of the rails curling up and forming "snake ends;" these would become detached from the wooden stringers, and sometimes would pierce the floors of the cars, injuring and maiming the passengers. The English DOUBLE-HEADED RAIL is supposed to possess an advantage over the American rail, from the fact that when one head is worn out it can be reversed, and the wheels can be equally accommodated by the other head. If such were really the case, its superiority would be unquestioned, but in actual practice it appears that, by the time one head is worn out, the other has received such injuries from hammering away in the iron chairs, that notches are found in the rail at such points as to render the surface unfit for the wheel to run on, and also to seriously injure the strength of the rail, and again, in many of the sections of English rails the form of the two heads is not alike, so this idea of reversing the rail does not seem to be of much importance. The STEEL-TOP RAIL, MITRE-JOINT RAIL, COMBINATION RAIL, all possess some merit, but their use has been extremely limited, and the writer deems it unnecessary to go into details respecting them. The BESSEMER STEEL RAIL, made by the Bessemer process, is now the standard rail, and as long as it can be purchased at its present low price, its advantages over the iron rail seem to admit of no argument. In fact the superiority of steel over iron rails is now no longer disputed; they are not only more durable, but are much stronger for the same amount of material, their comparative strength being in the same proportion as 5 to 3. Steel rails are less fibrous than iron, and consequently less liable to splinter off from use. By actual test, one steel rail has outworn 17 iron rails, with only $\frac{5}{16}$ of an inch worn off the top. Great care is necessary in selecting irons for the Bessemer process; only those containing very little sulphur and phosphorus can be used, the former causing red-shortness, and the latter cold-shortness, or brittleness. At the furnace of the Penna. Steel Works, a very superior iron is made which they themselves use in making steel rails. The ores used are chiefly magnetic, from Dillsburg, York County.

8

degree of hardness; this addition of speigel
to the metal produces a violent action,
which soon ceases, and the steel is then
poured into a ladle, from which it is sub-
sequently run into cast-iron moulds.   A
small test-ingot is taken from each charge,
and chemically tested.   The capacity of
a Bessemer plant is from 25 to 30 blows
per day.   From the moulds the steel
ingot is taken to the blooming mill, and
the ingots are reheated and rolled into
blooms, which are in turn heated again,
and then rolled into rails.   Great care is
necessary in manufacturing steel rails to
avoid brittleness, and to insure toughness.
When steel rails were first manufactured,
there was an uncertainty in the quality
of the material which raised quite a pre-
judice in the minds of many Engineers
against their use, but the process of
manufacture has now been brought to
such a high state of perfection that this
doubt no longer exists, and where the
traffic of the line will permit they are

generally adopted. The cost of steel rails at the present writing (1897) is about $20 per ton at the mills, or about $1880 per mile of sixty-pound steel.

The first steel rail was made in 1857, by Mushet, at the Ebbow-Vale Iron Company Works, in South Wales. It was rolled from cast blooms of Bessemer steel, and laid down at Derby, England, and remained sixteen years, during which time 250 trains, and at least 250 detached engines and tenders passed over it daily. Taking 312 working days in each year, we have the total of 1,252,000 trains, and 1,252,000 detached engines and tenders, which passed over it from the time it was first laid before it was removed to be worked over. Two steel rails of 21 feet in length were laid on the 2d of May, 1862, at the Chalk Farm Bridge, side by side with two ordinary iron rails. After having outlasted 16 faces of the ordinary rails, the steel ones were taken up and examined, and it was found that at the expiration of three years and three months

the surface was evenly worn to the extent
of only ¼ of an inch, and to all appear-
ance they were capable of enduring a good
deal more work. These two rails, during
a period of little more than three years,
had been exposed to a traffic of 9,550,000
engines,   trucks,   and   carriages,   and
95,577,240 tons, an amount of traffic
equal to nearly *ten* times that which destroyed the Great Northern iron rails in
three years' time.

In England rails are rolled from 15 to
21 feet long, the latter being the most
common size. In this country it is not
an uncommon feat to roll a perfect rail
60 feet in length, weighing 60 to 90 pounds
to the yard. A rail is generally from 3
to 4½ inches high (usually 4½ inches),
and the width from 2¼ to 3 inches at the
head, 3 to 5 inches on the flange, and
having a web or neck of ½ to 1 inch in
thickness. The life of *iron rails* of best
quality has been found to be 35,000,000
tons over a double line, or 17,500,000
tons over each single rail, and many from

the best makers stand only 5,500,000 to 15,000,000 tons, or equal to 100,000 trains of 150 tons each, independently of the length of time of the traffic. The wear may be estimated as the $\frac{1}{100000}$th part of the value of the rail each time a train goes over it; and if the value of a mile of iron be taken at $5000, the wear would be 5 cents per train per mile.

A great many American railroads are using a rail of the average section of 60 pounds per yard, some as high as 90.

Table giving the Number of Tons of Rails required to Lay One Mile of Single Track Railroad of different Weights of Rails, of Steel or Iron.

Weight of rail p'r yard.	Tons per mile.	Weight of rail p'r yard.	Tons per mile.	Weight of rail p'r yard.	Tons per mile.
8 lbs.	$12\frac{1280}{2240}$	45 lbs.	$70\frac{1600}{2240}$	64 lbs.	$100\frac{1280}{2240}$
12 "	$18\frac{1920}{2240}$	50 "	$78\frac{1280}{2240}$	65 "	$102\frac{320}{2240}$
16 "	$25\frac{320}{2240}$	52 "	$81\frac{1600}{2240}$	67 "	$105\frac{640}{2240}$
25 "	$39\frac{640}{2240}$	56 "	88	68 "	$106\frac{1920}{2240}$
30 "	$47\frac{320}{2240}$	57 "	$89\frac{1280}{2240}$	70 "	110
35 "	55	60 "	$94\frac{940}{2240}$	80 "	$125\frac{1600}{2240}$
40 "	$62\frac{1920}{2240}$	62 "	$97\frac{960}{2240}$	90 "	$141\frac{960}{2240}$

AN IRON RAIL is made by rolling together a number of separate pieces of iron which when placed in position preparatory to rolling, are called " RAIL PILES." These rail piles are formed in different ways according to the ideas of the manufacturer. The pile is first treated in a furnace to a welding heat, and hammered or rolled into a solid lump or bloom, which is again heated and rolled into the desired shape of rails. Formerly a very important fact to consider in deciding between iron and steel was, that after an iron rail became worn out and no longer fit for railroad service, it was still a marketable article, and could readily be disposed of as old iron, at at least two-thirds of its original cost, but what old steel rails were worth was a comparatively unknown quantity. This is no longer a consideration. Recent improvements in the melting of old steel have made that article a valuable asset, and old steel rails can readily be exchanged for new, plus a comparatively small cost for rerolling.

The following Table gives the Average Price, per Ton, of Iron and Steel Rails, New York, during the past Fifty Years.

Year.	Iron.	Steel.	Year.	Iron.	Steel.
1847	$70		1872	$90	$110
1848	60		1873	85	120
1849	50		1874	65	75
1850	45		1875	50	70
1851	45		1876	45	60
1852	45		1877	40	50
1853	75		1878	35	40
1854	80		1879	40	48
1855	60		1880	50	68
1856	60		1881	47	60
1857	65		1882	45	48
1858	50		1883	40	37
1859	50		1884		30
1860	45		1885		28
1861	40		1886		34
1862	36		1887		37
1863	70		1888		35
1864	153		1889		30
1865	84		1890		32
1866	80		1891		30
1867	80		1892		30
1868	78	$175	1893		30
1869	75	150	1894		28
1870	75	130	1895		28
1871	70	95	1896		28

New steel rails are worth only $20 per ton (the price recently fixed for 1897), and this low price, together with their acknowledged superiority over iron rails, has practically driven the latter out of the market. Nearly 90 per cent. of the track of all railways in this country is now of steel. It is not always best to buy a heavy rail; very often a lighter rail will do as much service for the amount of traffic, making quite a saving in the cost. A rail will usually wear out first at the ends, owing to the wheels hammering over the open joints, which are caused by the rails contracting and expanding at the different degrees of temperature. Many devices have been invented to overcome THE OPEN JOINT, but for all practical purposes they have not been successful; the joint still exists, and is a constant source of annoyance and expense; the only thing to be done is to make that joint as secure as possible by using good fastenings, and when laying track to make allowances for the contraction and expansion of the rails.

Table giving the Number of Rails and Joints per Mile of Single Track Railroad, for Rails of different Lengths.

Rails 24′ long each.	440 complete joints and rails.
" 25 " "	422 " " "
" 26 " "	406 " " "
" 27 " "	391 " " "
" 28 " "	377 " " "
" 30 " "	352 " " "

In order to fasten two rails together at their ends, some fastening or other coupling must be used. Time and space will not permit even mention by the writer of the hundreds of devices presented to his notice for effecting this object, very many having claims to originality of design, but very little else. On the many miles of railroad now operating in this country, but very little difference will be observed in the manner of fastening or "FISHING" the rails; and the general tendency appears to be towards the adoption as a

Continuous rail joint.

124

STANDARD JOINT of two iron bars, one on each side of the rails and bolted through. These bars are made in many different shapes and sizes, but generally speaking are much alike, simply two bars or plates ranging from 18 to 24 inches in length, and having holes punched in them for the bolts to go through, these holes corresponding in size and position to the holes in the rails. When the joints happen so that a cross-tie is immediately under it, then it is called a "SUPPORTED JOINT," but when the cross-ties come on each side of the joint, it is called a "SUSPENDED JOINT." The latter is undoubtedly the best, possessing greater elasticity and preserving the life of the rail by relieving the anvil pounding it would receive if too rigidly supported. FISH PLATES are usually quoted by the pound or per joint of two bars. The price per pound is about two cents, and per joint, the price will vary according to the weight and size of the bar. The following table has been

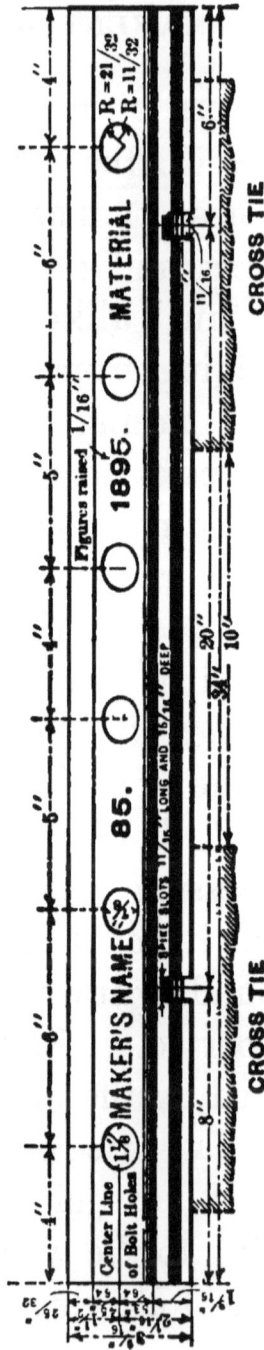

Fish plate, Pennsylvania Railroad standard.

prepared for the different sizes, weights, and present prices per pound and joint for

Plain Fish Bars.

(Original.)

Weight of rail.	Length of plate.	Weight of plates.	Price per pound.	Price of plate.	Joint with bolts.
30 lbs.	16½ n.	6½ lbs.	2 cts.	13 cts.	36 cts.
40 "	22 "	13 "	2 "	26 "	62 "
50 "	22 "	13 "	2 "	26 "	62 "
56 "	23 "	16 "	2 "	32 "	75 "
60 "	23 "	16 "	2 "	32 "	75 "
67 "	24 "	18 "	2 "	36 "	87 "
90 "	34 "	30 "	2 "	60 "	1.35 "

To connect the fish bars with the rails, 4 BOLTS are used (excepting the 34-inch plates, where 6 bolts are used), two in the ends of each rail; these bolts measure $\frac{3}{4}$ of an inch in thickness, and are of various lengths, usually, however, $4\frac{3}{4}$ inches long. The heads of the bolts are made square, oblong, or with round-button heads, the latter being the most common. The holes in one fish plate are made oval to fit an oval-headed bolt, and prevent it from turning round.

Rail joint, with fish bars, bolts, and nut locks.

The nuts are made square, or hexagonal. Bolts and nuts are usually quoted by the pound; a bolt and nut together will weigh about one pound. The price of Fish bolts and nuts is about 3¼ cents per pound, bolt and nut together.

Table of the Number of Fish Plates and Bolts required for One Mile of Single Track Railroad.

(Original.)

Length of rail.	No. of plates.	No. of bolts.	No. of joints.	Price per mile for complete joints.
			Complete.	Rail 60 lbs.
24 feet.	880	1760	440	$330.00
25 "	844	1688	422	316.50
26 "	812	1624	406	304.50
27 "	782	1564	391	293.25
28 "	754	1508	377	282.75
30 "	704	1408	352	264.10

The customary lengths of rails are seldom under 30 feet. The cost, therefore, of Fish plates and bolts for one mile of single track railroad, will be about $264, as shown by the preceding tables.

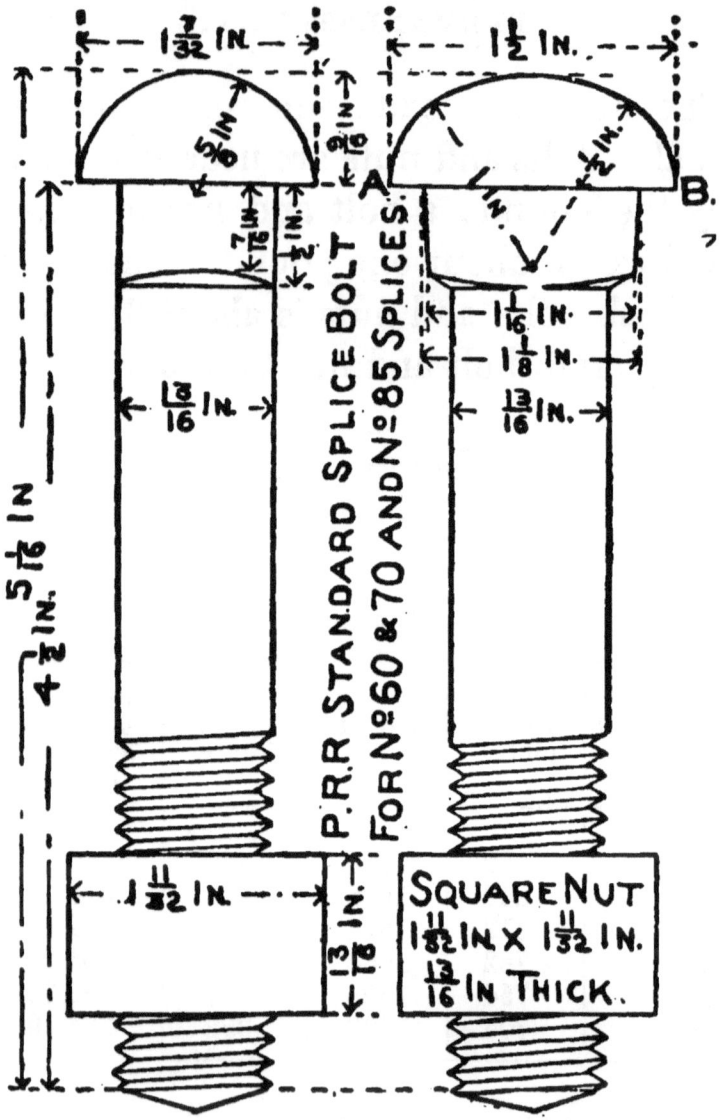

P.R.R. STANDARD SPLICE BOLT.
FOR Nº 60 & 70 AND Nº 85 SPLICES.

A.

B.

1 7/32 IN.

1 1/2 IN.

5/8 IN

9/16 IN

5/8 IN

1/2 IN

7/16 IN

1/2 IN

13/16 IN.

1 1/16 IN.

1 1/8 IN.

13/16 IN.

5 1/16 IN

4 1/2 IN.

1 11/32 IN.

13/16 IN.

SQUARE NUT
1 11/32 IN X 1 11/32 IN.
13/16 IN THICK.

180

The following tables contain all necessary information regarding bolts and
nuts, viz. :

Number to One Hundred Pounds.

	Square.	Hexagon.		Square.	Hexagon.
½ inch,	1100	1250	1 inch,	150	170
⅝ "	550	650	1⅛ "	98	110
¾ "	375	415	1¼ "	70	80
⅞ "	230	155	1½ "	45	55

Weight of Nuts and Bolt Heads in lbs.

Diameter of bolt in inches,	¼	⅜	½	⅝	¾	⅞
Weight of hexagon nut & head	.117	.057	.128	.267	.43	.73
Weight of square nut and head	.021	.069	.164	.320	.55	.88

Diameter of bolt in inches	1	1¼	1½	1¾	2	2½	3
Weight of hexagon nut & head	1.10	2.14	3.78	5.6	8.75	17	28.8
Weight of square nut and head	1.31	2.56	4.42	7.0	10.5	21	36.4

Standard for Screw-threads, Bolt-heads, and Nuts.

Adopted by the Master Car-Builders' Association.

Diameter of bolt.	No. of thr'ds per inch.	Diameter of bolt.	No. of thr'ds per inch.	Diameter of bolt.	No. of thr'ds per inch.	Diameter of bolt.	No. of thr'ds per inch.
$\frac{1}{4}$	20	1	8	$2\frac{1}{4}$	$4\frac{1}{2}$	$4\frac{1}{2}$	$2\frac{3}{4}$
$\frac{5}{16}$	18	$1\frac{1}{8}$	7	$2\frac{1}{2}$	4	$4\frac{3}{4}$	$2\frac{5}{8}$
$\frac{3}{8}$	16	$1\frac{1}{4}$	7	$2\frac{3}{4}$	4	5	$2\frac{1}{2}$
$\frac{7}{16}$	14	$1\frac{3}{8}$	6	3	$3\frac{1}{2}$	$5\frac{1}{4}$	$2\frac{1}{2}$
$\frac{1}{2}$	13	$1\frac{1}{2}$	6	$3\frac{1}{4}$	$3\frac{1}{2}$	$5\frac{1}{2}$	$2\frac{3}{8}$
$\frac{9}{16}$	12	$1\frac{5}{8}$	$5\frac{1}{2}$	$3\frac{1}{2}$	$3\frac{1}{4}$	$5\frac{3}{4}$	$2\frac{3}{8}$
$\frac{5}{8}$	11	$1\frac{3}{4}$	5	$3\frac{3}{4}$	3	6	$2\frac{1}{4}$
$\frac{3}{4}$	10	$1\frac{7}{8}$	5	4	3		
$\frac{7}{8}$	9	2	$4\frac{1}{2}$	$4\frac{1}{4}$	$2\frac{7}{8}$		

The distance between the parallel sides of a bolt-head and nut for a rough bolt shall be equal to one and a half diameter of the bolt, plus one-eighth of an inch.

The thickness of the heads for rough bolts shall be equal to one-half of the distance between their parallel sides.

The thickness of the nut shall be equal to the diameter of the bolt.

The thickness of the head for a finished bolt shall be equal to the thickness of the nut.

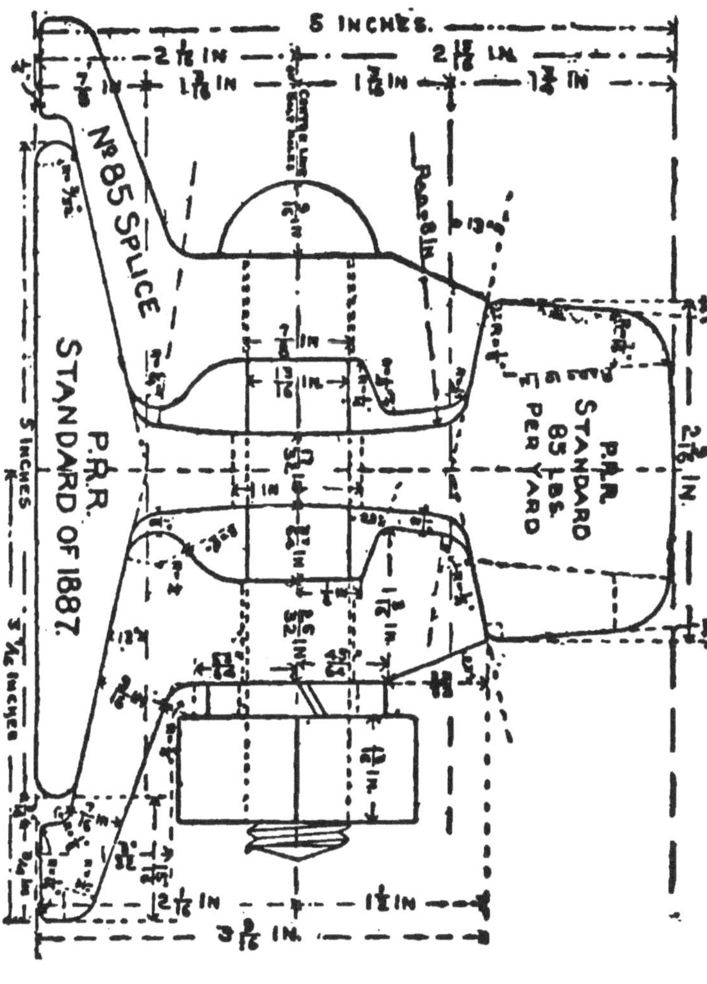
5 INCHES.
No 85 SPLICE
P.R.R.
STANDARD OF 1887.
P.R.R.
STANDARD
85 LBS.
PER YARD

The distance between the parallel sides of a bolt-head and nut and the thickness of the nut shall be one-sixteenth of an inch less for finished work than for rough.

Now, although the fish plates and bolts properly secured to the rails will connect them together lengthways, something is necessary to prevent them from moving sideways; this is done by using " HOOK-HEADED" SPIKES, which are driven into the sills or cross-ties close to the flange of the rail until their hooked heads, projecting over the flanges, will hold the rail firmly in position. SPIKES are of various sizes, styles, and dimensions, but it has been found from practice and repeated experiments that the plain square spike, $5\frac{1}{2}'' \times \frac{9}{16}''$, pointed like a wedge (say double the thickness of the spike), and driven squarely across the grain of the wood, is better than any of the numerous " ragged" or " spiral" devices. The STANDARD SPIKE for a broad-gauge (4' $8\frac{1}{2}''$) railroad

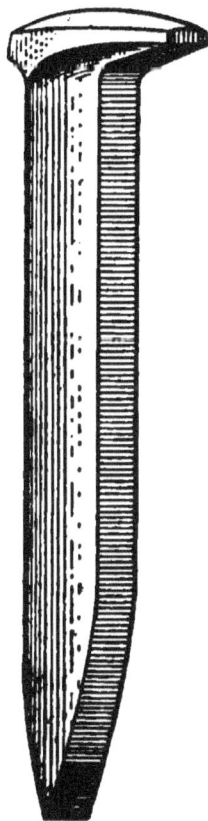

Standard railway spike

$(5\tfrac{1}{2}'' \times \tfrac{9}{16}'')$.

135

should measure $5\frac{1}{2}$ inches from the tip of the point to the under side of the head, and should be $\frac{9}{16}$ of an inch square in thickness. FOR NARROW-GAUGE (3 ft.) railroads, $4\frac{1}{2}$ inches by $\frac{7}{16}$ of an inch will do very well, and should be well driven, four to each cross-tie. Spikes are quoted by the pound, the present price being about $2\frac{1}{2}$ cents per pound; they are packed in kegs of 150 pounds each.

Railroad Spikes.

Size, measured under head.	Average No. per keg of 150 lbs.	Ties two feet between centres, four spikes per tie, makes per mile.	Rail used, weight per yard.
Inches.		Pounds. Kegs.	
$5\frac{1}{2}$ x $\frac{9}{16}$	280	5670=38	45 to 70
5 x $\frac{9}{16}$	300	5170=35	40 to 56
5 x $\frac{1}{2}$	340	4660=31	35 to 40
$4\frac{1}{2}$ x $\frac{1}{2}$	400	3960=27	30 to 35
4 x $\frac{1}{2}$	450	3520=24	28 to 35
$4\frac{1}{2}$ x $\frac{7}{16}$	510	3110=21 ⎫	25 to 30
4 x $\frac{7}{16}$	540	2940=20 ⎬	
$3\frac{1}{2}$ x $\frac{7}{16}$	675	2350=18 ⎫	20 to 25
4 x $\frac{3}{8}$	760	2090=14 ⎬	
$3\frac{1}{2}$ x $\frac{3}{8}$	890	1780=12 ⎫	16 to 20
3 x $\frac{3}{8}$	930	1710=$11\frac{1}{3}$ ⎬	

The foregoing TABLE gives the number of spikes to a keg, and the number of pounds and kegs to the mile, for different sizes; from which we ascertain that a mile of single track railroad will require (using the average weight of rail, 60 pounds) 38 kegs, or 5670 pounds of $5\frac{1}{2}'' \times \frac{9}{16}''$ spikes, costing $141.75. The smaller spikes are somewhat more expensive than the larger ones, by about $\frac{1}{2}$ cent per pound. A very important feature in the construction of a railroad is involved in TRACK-LAYING, and when any trouble occurs with the joints and fastenings it can generally be traced to poor or insufficient ballasting. The following contract for track-laying is expressed in the usual form of such agreements.

Contract for Track-laying.

Articles of Agreement, made and concluded this —— day of —— 18—, by and between —— of the first part and the

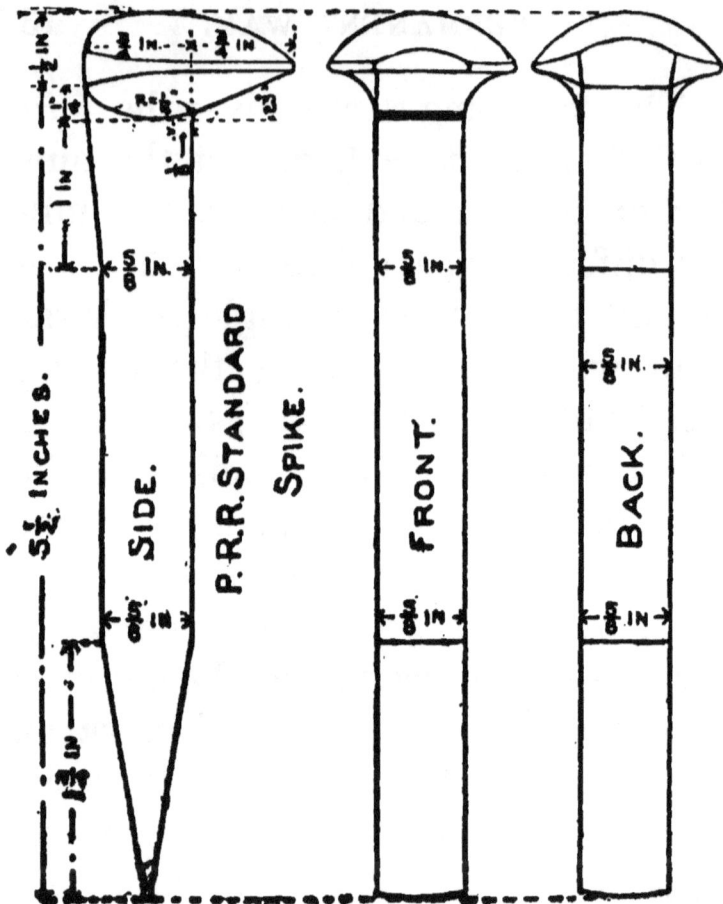

P.R.R. STANDARD SPIKE.

SIDE.   FRONT.   BACK.

138

—— Railroad Company of the second part.

WITNESSETH: That for, and in consideration of the payments and covenants, hereinafter mentioned, to be made and performed by the said party of the second part, the said party of the first part doth hereby covenant and agree to do all the Track-laying and Back-filling of the main track and sidings of the said —— Railroad Company, from —— to —— in accordance with the following specifications; and in conformity with the directions of the Engineer in charge, and to his satisfaction and acceptance; and to complete the same on or before the —— day of —— 18—.

SPECIFICATIONS.

The ties shall be laid accurately to the stakes as given by the Engineer in charge; they shall be carefully mauled down on, or into, the ballast, to such depths as the Engineer may direct, and in such a manner as to give them a firm, continuous, and even bearing thereon. Care shall be taken that all ties with rind be so adzed off as to present even and parallel bearing-surfaces for the rails to rest on. The ties shall generally be laid at the rate of twenty-six hundred and fifty to the mile, but the Engineer may increase or diminish the number at such points as he may consider necessary. Particular attention must be given to the ties immediately next the joints, that they be firmly bedded into, and have a continuous bearing upon the ballast, the largest being selected for this purpose. The rails shall be accurately laid to the line, and level-stakes given by the Engineer or his Assistants, and on all curves they must be bent to the proper curvature before being laid on the ties; they shall be laid to a gauge of ——, and so as to break joints. Care shall be taken that the proper space be allowed between the ends of the rails to provide for expansion, and, on curves, that the proper elevation, as fixed by the Engineer, be given to the outside rail. Each rail shall be securely spiked by two spikes, one on each side, to each tie; the spikes shall be well driven home, so as to bring the rails firmly down upon the ties; a greater or less number of spikes shall be used where the Engineer may require it. On Bridges the rails shall be spiked down upon the Track-stringers so as to give them a continuous bearing along their entire length, with such a number of spikes as the Engineer may direct.

The Sidings shall be laid at such points and of such lengths as the Engineer may direct, and of either new or old rails as he may deem expedient. In putting in the Switches and Frogs, care shall be taken to put them accurately to the position as determined by the Engineer; they must be laid upon ties specially provided for that purpose, which shall be so laid as to have a firm and continuous bearing upon the ballast. Each Frog shall be accurately laid with a uniform bearing upon four ties. All the backfilling that may be required shall be furnished by party of the first part, and shall consist of stone, furnace cinder, or gravel, of a size to pass through a ring of three inches diameter. After the track is firmly and accurately laid, it shall be properly surfaced, tamped, and lined up; and the spaces between the ties shall be filled with good broken stone or cinder as above specified (and as directed by the Engineer, not exceeding four hundred cubic yards to the mile) to the top and for the whole length of the ties. All material that shall have been thrown into the ditches, by the party of the first part, shall be removed therefrom, and the road properly ditched and cleaned up.

The party of the first part shall maintain and keep the track in good repair until the same is accepted by the Engineer; and no length of track shall be accepted and taken off the hands of the party of the first part, except at the option of the Engineer, until the whole shall have been completed.

And, the party of the first part doth further agree to load upon cars, to be furnished by the party of the second part, and transport all iron, chains, spikes, ties, and other material that may be required for said track-laying from such points along line of said railway at which they may be piled, and to unload the same from said cars at the nearest accessible point to the place where the track-laying is in progress; said loading, transporting, and unloading to be done promptly by the party of the first part and at his own expense. The party of the first part will be required to insert such small open-plank cross-drains under the railroad or crossing roads, or at such other points as the Engineer may direct; said drains to be paid for by the party of the second part at the valuation of the Engineer.

All the work that may be required to complete the track ready for the running of trains shall be done by the party of the first part, if required by the Engineer, and be paid for by the party of the second part at the valuation of the Engineer; but no work shall be done except upon the orders or instructions of the Engineer, and no claims for extra work will be allowed, unless said work was done under the orders of the Engineer, and the claim presented on or before the first day of the month next after such work was done. No extra allowance will be made for any delays that may occur in the construction of the work, but it may entitle the party of the first part to an extension of time for completing the work sufficient to compensate for the detention; to be determined by the Engineer. And the said party of the second part doth further agree and promise to pay to the said party of the first part for completing this contract as follows, viz.:—

For each and every mile of main track and siding, —— dollars.

For each and every cubic yard of back-filling, —— dollars.

For each and every Frog and Switch that may be set complete, —— dollars.

On or about the last day of every month, during the progress of the work, an estimate shall be made of the relative value of the work, to be judged by the Engineer, and upon his certificate of the amount being presented to the said party of the second part, or such disbursing agent as they may appoint, —— of the amount of said estimate shall be paid, in current funds, to the party of the first part between the tenth and twentieth of the ensuing month. And when all the work embraced in this Contract is completed in accordance with the specifications, and to the satisfaction and acceptance of the Engineer, there shall be a final Estimate made of the quality, character, and value of said work agreeably to the terms of this agreement, when the balance appearing due to the said party of the first part shall be paid to —— or ——, giving release under the seal of the said —— Railroad Company from all claims or demands whatsoever growing in any manner out of this agreement.

IT IS FURTHER covenanted and agreed between the said parties, that the said party of the first part shall not let or transfer this contract, or any part thereof, to any other person without the written consent of the party of the second part, but shall give —— personal attention and superintendence to the work. And it is further understood that the Engineer shall have the right to regulate, from time to time, the wages of labor upon the line of the work so as to maintain a proper distribution of the force, and prevent the injurious effects of competition among the contractors for hands.

IT IS FURTHER agreed and understood that the work embraced in this contract shall be commenced on or about the — day of ——, 18—, and prosecuted with such force as the Engineer shall deem adequate to its completion within the time specified; and if at any time the said party of the first part shall refuse or neglect to prosecute the work with sufficient force, in the opinion of the Engineer, for its completion within the time specified in this agreement, then in that case the Engineer, or such agent as he may appoint, may proceed to employ such a number of workmen, laborers, and overseers as may, in the opinion of the Engineer, be necessary to insure the completion of the work within the time hereinbefore specified, at such wages as he may find it expedient or necessary to give, pay all persons so employed, and charge over the amount so paid to the party of the first part as for so much money paid to said party of the first part in this contract; or said Engineer may at his discretion, for failure to prosecute the work with an

adequate force, for non-compliance with
his directions in regard to the manner of
completing it, or for any other omission
or neglect of the requirements of this
agreement and specifications on the part
of the party of the first part, declare this
contract or any portion of it forfeited;
which declaration and forfeiture shall exonerate the said party of the second part
from any and all obligations and liabilities
arising under this Contract, the same as
if this agreement had never been made;
and the reserved percentage of —— upon
any work done by the party of the first
part may be retained forever by the party
of the second part; and it is mutually
agreed and distinctly understood that the
decision of the Chief Engineer of the said
—— Railroad Company shall be final and
conclusive, in any dispute which may
arise between the parties to this agreement, relative to or touching the same,
and —— said party of the first part doth
hereby waive any right of action, suit or

suits, or other remedy at law or otherwise, by virtue of said covenant, so that the decision of the said Engineer shall, in the nature of an award, be final and conclusive on the rights and claims of said parties.

IN WITNESS WHEREOF, the President of —— Railroad Company hath signed the same and caused the corporate seal of the said company to be attached, and the said —— hath hereunto set — hand and seal the day and year first above written.

Attest: —— ——,
Secy.

{ Seal of
Company. }

—— ——,
President.

—— ——. [SEAL.]
—— ——. [SEAL.]
—— ——. [SEAL.]

WITNESS:

—— ——.

—— ——.

In making a contract for building a railway, it is customary, when the Engineer makes his monthly statement or estimate, to reserve 10 per cent. of it until the final estimate is made, or until the work has been accepted by the Engineer, at which time it is paid over to the contractor. Great attention should be given to tracklaying, and in no instance should a contractor undertake a contract for tracklaying who is not familiar with the process or operation which he undertakes. Ignorance of the subject will be a cause of heavy losses by the contractor and endless annoyance to the Engineer and Railroad Company. A GANG OF ABOUT 40 MEN is as large a party as can work advantageously together, and at one point. This gang is sufficiently strong in numbers to lay about three-quarters of a mile of track per day. The entire COST OF LAYING THE TRACK, including loading and unloading, and transporting the material, together with all backfilling and surfacing, will be on a single track railroad about $500 per mile. This also includes the laying of frogs and switches, curving of rails, and all incidental expenses properly belonging to the laying of the track.

VALUE OF IRON *per Gross Ton at from 1-10th of a cent to 10 cents per lb., increasing at the rate of 1-10th cent per lb.*

Per lb. in cts.& 1-10ths.	Price per ton.	Per lb. in cts.& 1-10ths.	Price per ton.
1-10	$2 24	2	$44 00
2-	4 48	1-10	47 04
3-	6 72	2-	49 28
4-	8 96	3-	51 52
5-	11 20	4-	53 76
6-	13 44	5-	56 00
7-	15 68	6-	58 24
8-	17 92	7-	60 48
9-	20 16	8-	62 72
1	22 40	9-	64 96
1-10	24 64	3	67 20
2-	26 88	1-10	69 44
3-	29 12	2-	71 68
4-	31 36	3-	73 92
5-	33 60	4-	76 16
6-	35 84	5-	78 40
7-	38 08	6-	80 64
8-	40 32	7-	82 88
9-	42 56	8-	85 12
		9-	87 36

Per lb. in cts.& 1-10ths.	Price per ton.	Per lb. in cts.& 1-10ths.	Price per ton.
4	$89 60	7	$156 80
1-10	91 84	1-10	158 04
2-	94 08	2-	161 28
3-	96 32	3-	163 52
4-	98 56	4-	165 76
5-	100 80	5-	168 00
6-	103 04	6-	170 24
7-	105 28	7-	172 48
8-	107 52	8-	174 72
9-	109 76	9-	176 96
5	112 00	8	179 20
1-10	114 24	1-10	181 44
2-	116 48	2-	183 68
3-	118 62	3-	185 92
4-	120 96	4-	188 16
5-	123 20	5-	190 40
6-	125 44	6-	192 64
7-	127 68	7-	194 88
8-	129 92	8-	197 12
9-	132 16	9-	199 36
6	134 40	9	201 60
1-10	136 64	1-10	203 84
2-	138 88	2-	206 08
3-	141 12	3-	208 32
4-	143 36	4-	210 56
5-	145 60	5-	212 80
6-	147 84	6-	215 04
7-	150 08	7-	217 28
8-	152 32	8-	219 52
9-	154 56	9-	221 76
		10	224 00

TRESTLES are wooden supports designed to carry the roadbed of a railway

over any depressions or water-courses, or
other places where it is required to avoid
the expense of embankments, as, for
example, where earth cannot be obtained
for grading and filling in. They are frequently used on railroads, as they cost
only about one-half the price of embankments. Very often they are used
only as a temporary expedient, to be filled
in with earth as soon as the road has been
completed, and is in operation. The design of a trestle varies with the height it
is proposed to make it. The approximate
COST OF TRESTLING, where the height is
not more than 30 feet, is about $6 per run-
ning or lineal foot. The timber used in
their construction is of the kind to be
found along the line or contiguous to it.
Over swamps and marshy grounds a
cheap and economical foundation can be
secured by driving piles, as in the annexed engraving, and capping them with
stout pieces of timber roughly hewn to
shape. These caps are made to support

the longitudinal stringers, which in turn support light plank-sills carrying the rails. For a single track railroad not

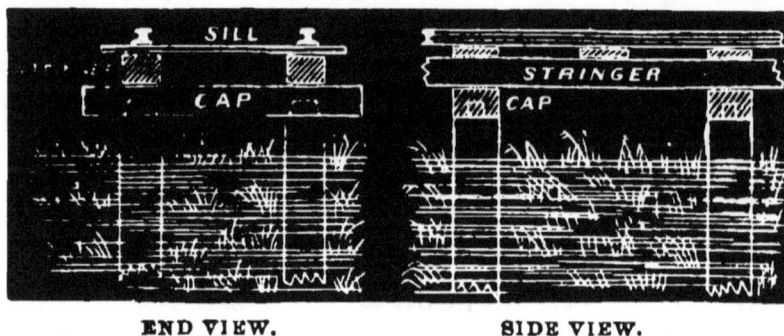

END VIEW.   SIDE VIEW.

Fig. 6.

having any extraordinarily heavy traffic, the timbers forming this trestle may be of the following dimensions:—

Piles . . .	12″	diameter.
Cap . . .	12″x14″	"
Stringers . .	12″x14″	
Sills . . .	3″ planking.	

And the entire cost in position will not be more than $2.50 per lineal or running foot, provided the timber can be procured along the line of the road. If necessary, light trestling 10 or 12 feet high can be erected on the caps with safety. Care should be taken to drive the piles so as to come directly under the line of each rail, and the cap should be neatly mortised to the pile. Longitudinally the piles may be driven from 8 to 10 feet apart. The gauge of the road should determine the cross distances. On the Northeastern Railroad, in South Carolina, such a trestle has been in use for 25 years, and with careful watching and necessary repairs is still in use. Other roads in the South also use it; notably the Savannah and Charleston Railroad. White pine is generally preferred; yellow pine where it can be procured.

On the subject of BRIDGES the author of this work is limited to a mere notice; no abridgment of so important a subject would be at all satisfactory to an Engineer, and of very little use to the unscientific reader. It is, therefore, thought advisable to simply add a few tables of reference suitable for a handbook. There are so many excellent works on bridges that an inquiring reader can easily follow up the subject in detail for himself.

In this country iron is largely used in the construction of bridges. For approximating roughly the WEIGHT OF IRON BRIDGES we have the following from Trautwine, for spans not exceeding about 300 feet.

RULE.—(Span + sq. root of span) × 6.4 = weight *in pounds*. This means the weight of the two trusses only, as used for a single track railroad. The weight of cross girders of floor, lateral bracing, floor boards, rail strings, and rails, with their spikes, bolts, rods, nuts, washers, etc. etc., will not vary much per foot run of span, in spans exceeding about 50 feet. From this rule the following table is deduced:—

(TRAUTWINE.)

Table of approximate average weights per foot run of span, of only the two iron trusses together, of a strong single-track railway bridge of the ordinary systems. Also the weights including the floor, lateral bracing, etc., complete, taken at 18 tons, or 403 lbs. per feet run of span.

Clear span.	Weight of only the two trusses together; per foot run of span.		Weight of two trusses, road-way, etc., com-plete; per foot run of span.		Entire weight of bridge, and max. load; per foot run.
Feet.	Tons.	Lbs.	Tons.	Lbs.	Tons.
5	.021	47	.201	450	4.201
7½	.029	65	.209	468	3.709
10	.038	85	.218	488	3.218
12½	.046	103	.226	506	2.626
15	.054	121	.234	524	2.534
17½	.062	139	.242	542	2.342
20	.070	157	.250	560	2.150
25	.086	193	.266	596	1.966
30	.101	226	.281	629	1.881
35	.117	262	.297	665	1.797
40	.133	298	.313	701	1.713
45	.148	332	.228	735	1.628
50	.163	365	.343	768	1.543
60	.194	435	.374	838	1.374
70	.224	502	.404	905	1.404
80	.254	569	.434	972	1.434
90	.284	636	.464	1039	1.464
100	.314	703	.494	1106	1.494
110	.344	771	.524	1174	1.524
120	.374	838	.554	1241	1.554
130	.404	905	.584	1308	1.584
140	.434	972	.614	1375	1.614

Clear span.	Weight of only the two trusses together; per foot run of span.		Weight of two trusses, roadway, etc., complete; per foot run of span.		Entire weight of bridge, and max. load; per foot run.
Feet.	Tons.	Lbs.	Tons.	Lbs.	Tons.
150	.463	1037	.643	1440	1.643
160	.493	1104	.673	1507	1.673
170	.523	1172	.703	1575	1.703
180	.553	1239	.733	1642	1.733
190	.583	1306	.763	1709	1.763
200	.612	1371	.792	1774	1.792
225	.687	1539	.867	1942	1.867
250	.760	1702	.940	2105	1.940
275	.833	1866	1.013	2269	2.013
300	.907	2032	1.087	2435	2.087

WOODEN BRIDGES WEIGH about the same as iron ones of equal strength.

FOUNDATIONS for the abutments of bridges and culverts must receive careful attention. When the surface of the ground upon which it is intended to build masonry is hard and compact for a distance or depth of 5 or 6 feet, no extra precautions need be taken beyond removing a foot or two of the surface before commencing the masonry; but in soft marshy ground it is necessary to drive piles and obtain a foundation, or else build cribs of rough hewn timber to sink to the hard soil by filling with stones. The Reading Railroad Company's wharves at the foot of Willow Street, Philadelphia, are built on piles driven to a depth of about 8 to 15 feet in the mud. The tops of these piles were, after being driven, cut off at low-water mark, and were then clamped together in rows, with $6'' \times 10''$ timbers bolted on each side of the pile (at low-water mark), which had been previously cut and shouldered, thus—

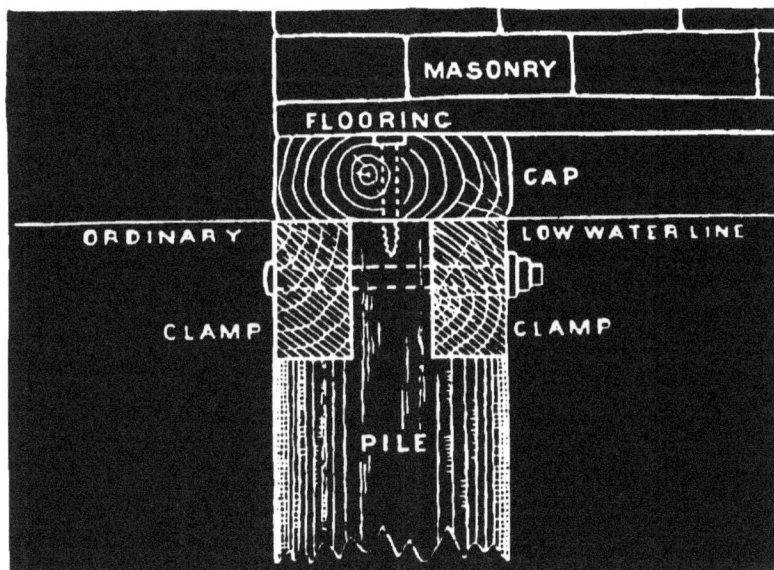

Fig. 6a.

A screw bolt passing through these clamps and the head of each pile held them firmly in position. On top of these clamps, as shown in the illustration, rested a cap of roughly hewn timber sufficiently large to cover both clamps and the heads of the piles. Across these caps the flooring was laid, and on that the masonry was built. The piles were driven in regular rows averaging about 3 feet apart, and for the most part were of good yellow pine, and were pointed with the axe before being driven. The "monkey" or hammer used in driving them weighed about 2700 pounds, and required on an average 20 blows to drive them home.

For the extreme end of the wharf a section of crib work was made, which after being towed into position, was sunk by throwing stones into it until the top of it came to low-water mark. Work was then resumed on it, and a flooring put in, on top of which more crib work was built, and then filled in with mud and dirt dredged from the bottom of the river and elsewhere. Mr. Wm. Rotan was the contractor who built these piers or wharves, acting under the immediate supervision of the writer.

CULVERTS are small openings placed in the bank at any depression in the ground, to permit the water to drain through the bank, and not become dammed up against it. An ordinary single box drain is built in the following manner: Stake out the four corners for a foundation, and dig out the inclosed space at least 12 inches deep. Then commence paving in this space with stones set on edge, taking care to have "headers" rammed down at each mouth a foot further or deeper than the paving. On top of this commence the walls, making each wall as thick as the drain is wide. Then place the roof or curbing in position, allowing each curb a foot rest on each wall, then back the remaining space on the wall with small stones or "spawls." For a DOUBLE BOX DRAIN

11

proceed as in single drain, the only difference being the middle wall which
should be one foot (at least) thick, allowing each curb to rest on it 6 inches, and a
foot on each side wall. DRAINS ON A SKEW
should be avoided by changing the course
of the stream. A DRAIN ON A CURVE
should be avoided altogether. Culvert
masonry will COST about $3.50 per cubic
yard.

CHAPTER V.

To enable an engine or train to pass safely from one track to another, an arrangement of rails and levers must be introduced, commonly termed a CROSS-OVER or TURNOUT. A SWITCH consists simply of two movable rails, essentially part of the main track, which can be moved so as to connect either with the main track or siding, and is made and called "RIGHT HANDED" or "LEFT HAND-ED," as the case may be. The word Switch comes from the German *sweig*, signifying *a branch* or *twig*. A right-handed switch is determined in this manner: Stand at the toe of the switch, or the joint made by the movable rails, and looking towards the frog; if the siding branches off to the right it requires a right-handed switch, but if it branches off to the left,

163

then a left-handed switch is necessary. This applies more particularly to the patent safety switches, not much difference being noticeable in the ordinary switches called stub switches. The ends of the movable rails of a switch, being the centre or fulcrum around which they move, are called the "HEELS," and the other ends of the movable rails are called the "TOES." The distance the toes move when the switch is being operated is called the "THROW," and is usually five inches. These movable rails, when set for the siding, form the tangents at their toes of which the turnout curve begins. The angle with which this curve crosses the main track rails determines the size or NUMBER OF THE FROG. The FROG DISTANCE is measured from the toe of the switch to the point of the frog. THE NUMBER OF A FROG is the proportion between the length and breadth of its point; that is, a point measuring six feet in length and one foot across its base would be a number six frog point. Frogs are

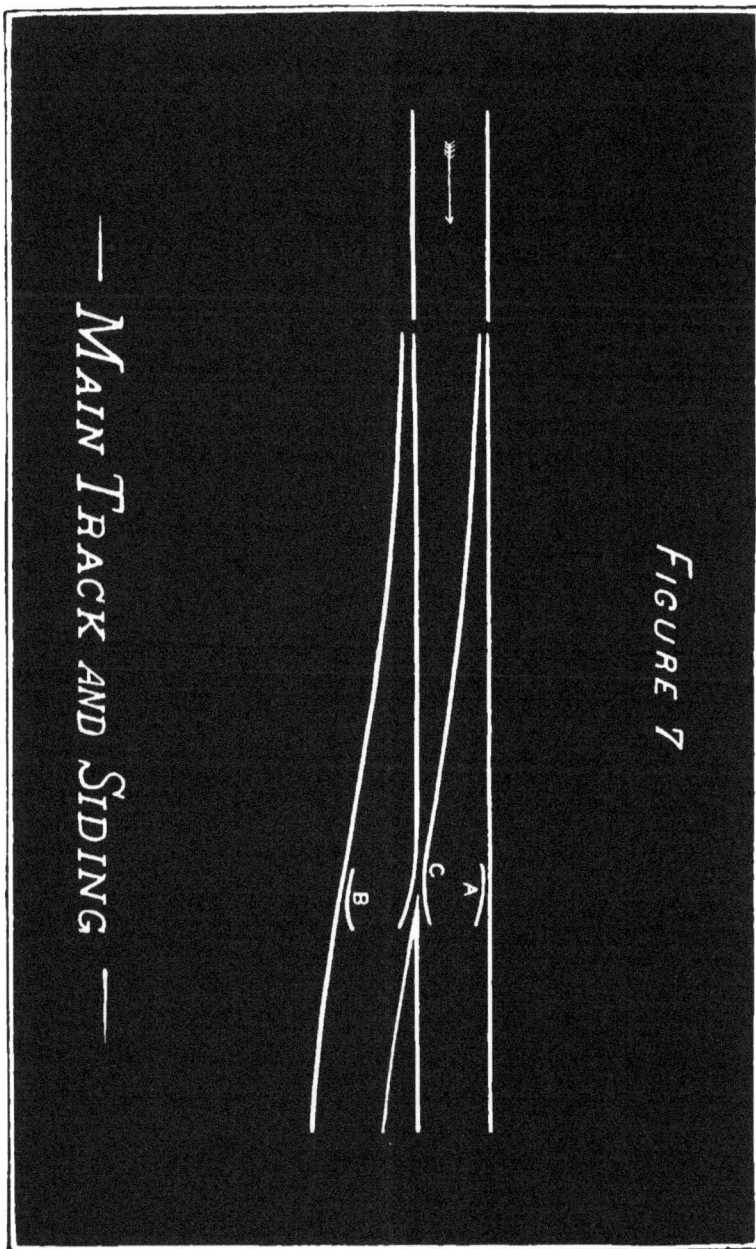

FIGURE 7

MAIN TRACK AND SIDING

A B, wing or guard rails ; C, frog.

numbered usually from four to twelve without any fractional sizes, although special angles are often required. The "THROAT" of a frog is the space between the wings and the point in which the flanges of the wheels run. The following table gives the frog distances for the different numbers of the frogs.

Table of Frog Distances.
(Original.)

Switch 30' long, throw 5".		Switch 26' long, throw 5".		Switch 24' long, throw 5".		Switch 20' long, throw 5".		Frog angle.
No. Frog.	Frog dist.	No. Frog.	Frog dist.	No. Frog.	Frog dist.	No. Frog.	Frog dist.	
12	87.9'	12	86.1'	...	...	...	...	4O.4O'
11	81.5	11	80.0	11	79.1'	11	76.7	5.02
10	75.0	10	73.7	10	72.7	10	70.8	5.44
9	63.0	9	67.2	9	66.5	9	64.7	6.21
8	61.0	8	60.5	8	60.0	8	58.5	7.10
7	54.3	7	53.6	7	53.0	7	52.0	8.10
6	47.1	6	46.6	6	46.2	6	45.4	9.32

THE STUB SWITCH.

The most simple form of switch in use

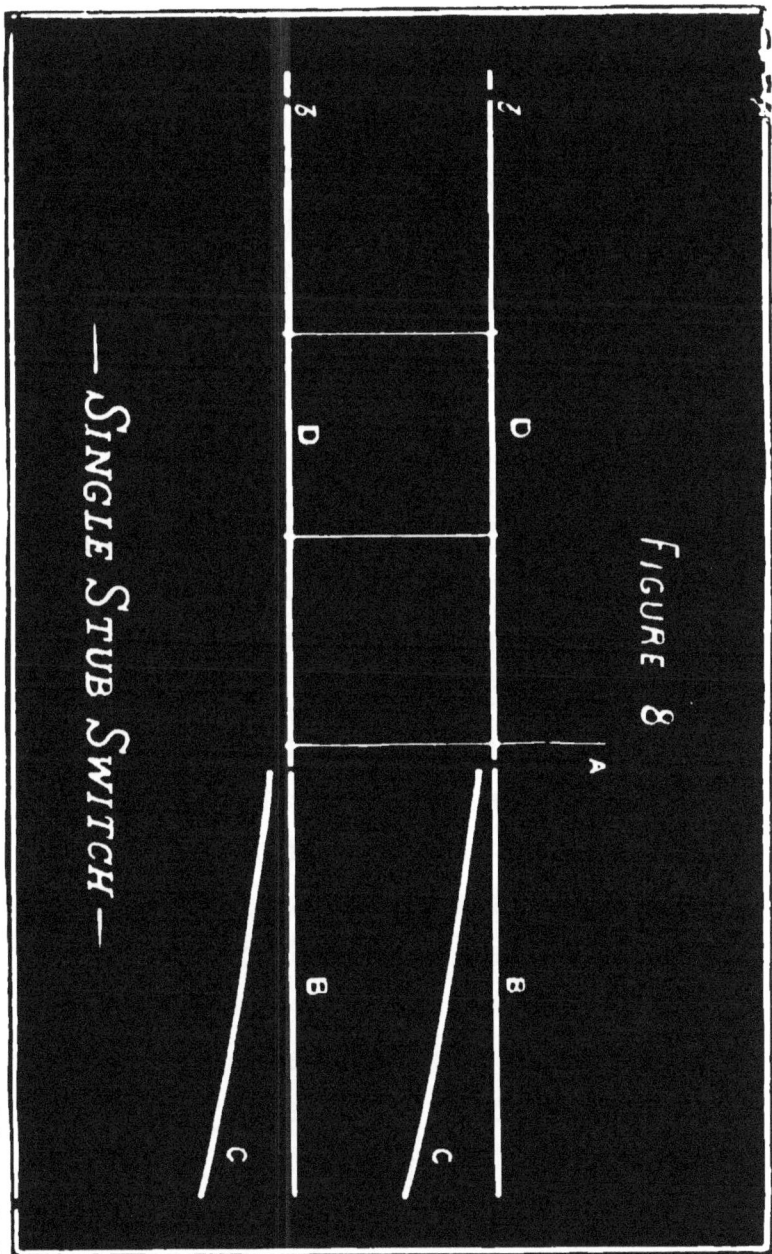

FIGURE 8.

— SINGLE STUB SWITCH —

A, switch lever; B B, main line fixed rails; C C, siding rails; D D, movable switch rails clamped together; b b, heels of switch.

is the common stub switch. It consists mainly of two movable rails D D—part of the main track—clamped together with "tie rods" so as to move together when thrown back or forth by means of any simple lever at A (fig. 8). The manner of working this switch is so simple as to require no explanation, though the minor details of construction admit of much variety. At the heels *b b* the rails are held firmly in position by means of fish plates. Track-men very often make this joint a "loose joint;" this is entirely unnecessary to effect the object intended; the movable rails should never be less than twenty feet long, and the spring of a rail of that length is amply sufficient for the throw of the switch, and is consequently all that is required. The toes of the switch should always rest on cast-iron "chairs," in order that the switch can be worked easily, and also to prevent the sharp ends of the rails from sinking into the wooden support or sill. Wrough iron straps should also be used

—fish plates do very well—at least three on each side, to be placed under the movable rails as sliding blocks. The chair is usually made of cast-iron to fit the end section of the two fixed rails, and to form a positive base for the movable rail to slide on, and should measure at least 12 × 16 inches, and have not less than one inch metal in its weakest part.

This chair is also made of wrought iron, with inverted clamps to hold the ends of the fixed rails, and sometimes of iron and wood together, forming a cushion and giving a certain amount of elasticity. A pair of these chairs will cost about $5.00.

The "tie rods" connecting the two movable rails are made in many different styles, all of which have some merit, except the kind which necessitates drilling the flanges of the rails. It has been proven by practice that any imperfection existing in the flange of a rail, more particularly of a steel rail, seriously affects the strength of the section, and is almost certain to cause fracture. A hole drilled or punched through the flange of a rail will generally produce the same result. A tie rod made of square iron $1\frac{1}{2}'' \times 1''$ with clamps on each end made to fit the lower half of the rail section is the very best in use. Five tie rods are generally used to each switch, and are placed five feet apart, the first one commencing ten inches from the toe of the switch. The "stand" can be made of wood or iron in any variety of shape or size. The best are made of cast iron, weighing about 150 pounds. When a "target" is used the stand should not weigh less than that, in order to have strength sufficient to resist the sudden wrenching and jarring produced in throwing the switch. Care should be taken when laying the switch to have the stand firmly secured with two or more screw bolts passing through the base of the stand and the sill it rests on, in addition to the ordinary number of spikes. And the connecting rod by means of which the stand lever is connected with the switch

Switch stand, with target

should be supplied with a sleeve-joint to admit of tightening up as it becomes loose by wear and length of service. Where the stand and target would take up too much room, the ground lever can be used to advantage. It is so familiar to all railroad men as to require no further mention. A "three throw," or, as it is often incorrectly termed, a "double throw" switch, does not differ materially from the single stub switch described above, except simply in the "chairs," which are made larger to allow the introduction of an additional rail. All switch stands are made for a "three throw" switch, and are used for both kinds. The stub switch has also been made practically a safety switch by means of additional castings so arranged and fastened to the switch that they receive the wheels of the engine, when they leave the track by accident, and guide them safely back again. There are several modifications of this principle, but in the main idea they are simi-

lar. At the best they are clumsy un-
mechanical contrivances, and can rarely
be depended on to do the work intended.
There are a very few exceptions, but the
first cost of these is in most cases too
great to admit of their use.

Although a comparatively safe switch
on our leading railroads, where they are
well protected by signals and other modern appliances, the stub switch is in itself
an imperfect, dangerous switch and open
to many serious objections, the principal
one being the noisy open joint causing
enormous wear and tear of the rolling
stock and destructive mashing of the rails.
Even by exercising the utmost care the
joints formed by the movable rails will
prove a source of endless annoyance, often
interlocking by expansion in summer,
and growing so large in winter by the
rails contracting and "crawling" as to
need frequent renewal. The cost of a
stub switch complete, including the chairs,
rods, stand, and target, ready for laying

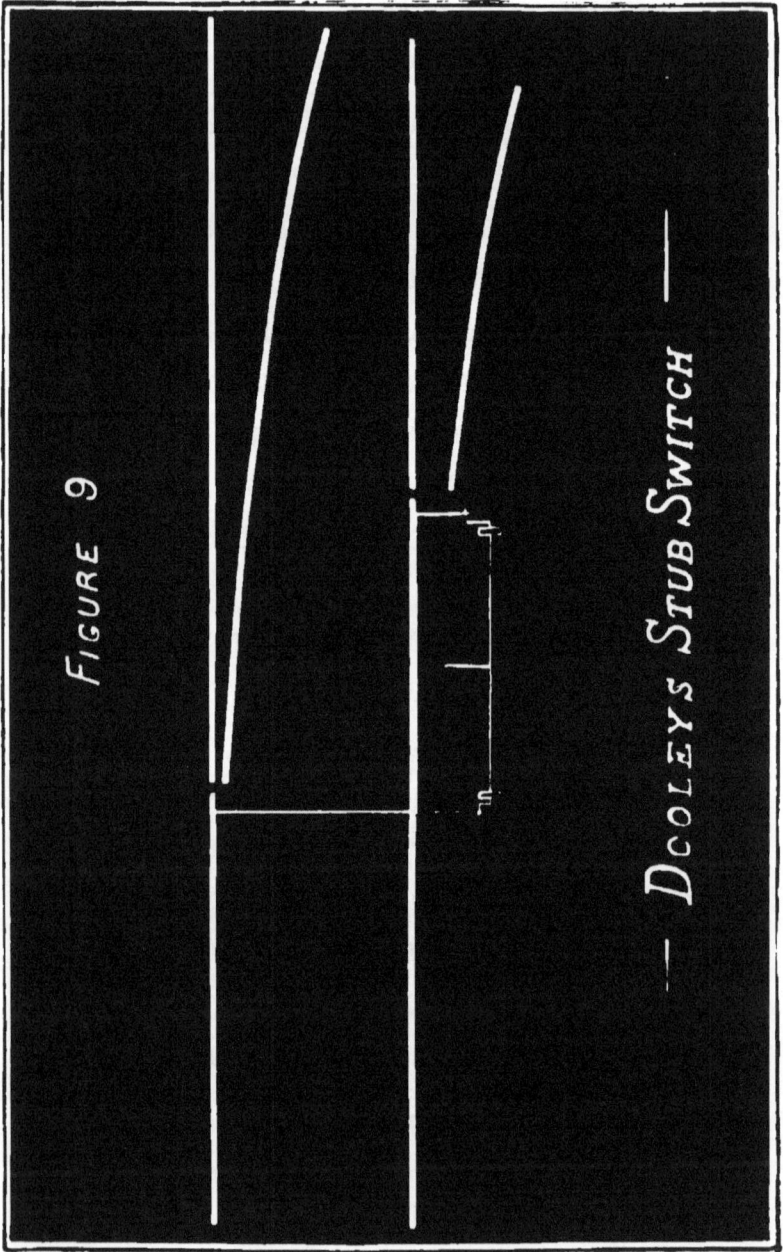

FIGURE 9

— DOOLEYS STUB SWITCH —

in the track, is about $35.00 per set. DOOLEY's SWITCH, by having the joints of the movable rails placed so that the wheels when riding over them will go over one joint at a time, is similar to the stub switch, except that the joints are not placed opposite to each other, one joint being a few feet back of the line of the other; otherwise it is a stub switch, and costs about the same. THE NICOLLS' SWITCH, designed by the author with the same idea—to lessen the jar and shock produced by the wheels passing over two open joints—has only one open joint in its construction.

As shown in the figure, the switch is set for the main line. A is one of the main track rails, unbroken and continuous; C and D form together the other main line rail, having an open joint at H, as in the ordinary stub switch. The rail B is planed or tapered to a point, and when closed fits up snugly to the rail A, as in the Lorenz Safety Switch (described

THE RAILWAY BUILDER.

hereafter). B and C are the two movable rails which, unlike other switches, move in *opposite* directions by means of the lever F and the two cranks G and H. At *b* and *c* these two movable rails are connected by fish plates with the fixed rails of the main track and siding, as described previously on page 141. To operate the switch it is only necessary to raise the handle F, and throw it over from the track as in the common ground lever switch; this motion by means of the crank H draws the rail C over in line with the rail E of the siding, and by means of the crank at G shoves the pointed rail B over against the rail A, thus setting the switch for the siding. The simplicity of its design, the fact of its having only *one* connecting rod across the track, and having only *one* joint, establishes its superiority over the stub switch, which has five connecting rods and two open joints.

A SAFETY SWITCH is so constructed that if it be left by accident set wrong, an engine or train will not be thrown from the track when attempting to pass through it. Very many devices have been invented to effect this object, and some merit can be claimed for almost all that are now in use. A favorite safety switch in England, and one which has given general satisfaction in this country also, is THE "SPLIT RAIL SWITCH," or, as it is better known in this country, the LORENZ SAFETY SWITCH, owing to the fact that Mr. Wm. Lorenz, Chief Engineer of the Reading Railroad Company, greatly improved its construction by the introduction of rubber springs. The construction of the switch is well defined by its name, the two movable rails being split, or planed down to points which fit up closely alongside of the outer rails of the track. The flanges of the wheels of the engine passing between these points and the outer rails shove them aside, and by doing so operate the switch. The rubbers which Mr. Lorenz introduced are placed in the

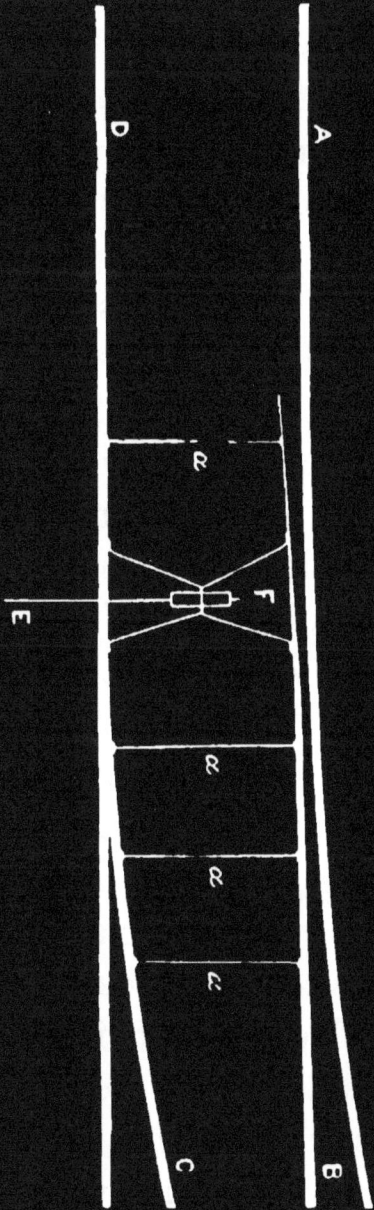

LORENZ SAFETY SWITCH—

FIGURE 11

rods which are used to operate the switch, and in such a manner that any strain coming on the rods is received by them, and owing to their elastic nature is rendered harmless. The split rail switch, as shown in the figure, is operated as follows: In the diagram the switch is shown set for the siding. A represents a continuous rail belonging to the side track, and D the continuous rail of the main track, both of which are unbroken rails, perfect in their entire length. This is one of the chief merits of the switch, as in either case, whether set for the siding or for the main line, one of the rails will be continuous and unbroken, and consequently no jarring or noise is experienced when running over it. B and C represent two rails which are planed and made pointed, so as to fit up closely to the rails A and D. These two pointed rails are clamped together by tie rods (usually five in number) so as to be thrown in the same direction and at the same time by means of any common lever at E. By means of this lever the switch is operated. It will readily be observed that if an engine should attempt to pass from right to left on the main track while the switch is set for the siding, as shown in the diagram, the flanges of the wheels will go between the rails D and C and force them apart, and the rail D being immovably spiked down to the sills will not move, so the pointed rail C must move, and also its opposite B, both being clamped together and only held in position by the connecting rod of the lever which will break or yield, allowing the flange to pass through, the engine consequently keeping on the track. In like manner the flanges of the wheels will operate against the rail B should a train attempt to pass *out* of the siding when the switch is set for the main track. To avoid this injury to the switch, that is, the breaking of the connecting rod, a spring has been introduced, indicated at F in the figure, which allows enough movement to make space enough for the flanges to pass between the rails, and which, after they have done so, by their elasticity draw the pointed rails back again into position for the main track. This spring Mr. Lorenz makes of rubber, and for many years has used it on the Reading Railroad. Some other roads, however, use coil springs of steel, which answer the purpose probably as well.

For a double track railroad where the trade is generally moving in one direction and on its own line of rails, this switch is unquestionably the best in use; many years of experience and trial have proved its superiority over other safety switches; but on a single track railroad where the switch would have to take the trade at both ends, the objection is raised of "running against the points." This switch can also be made a "three throw" switch, operated with two separate levers, A B, as shown in fig. 12. The cost of a Lorenz Safety Switch is about

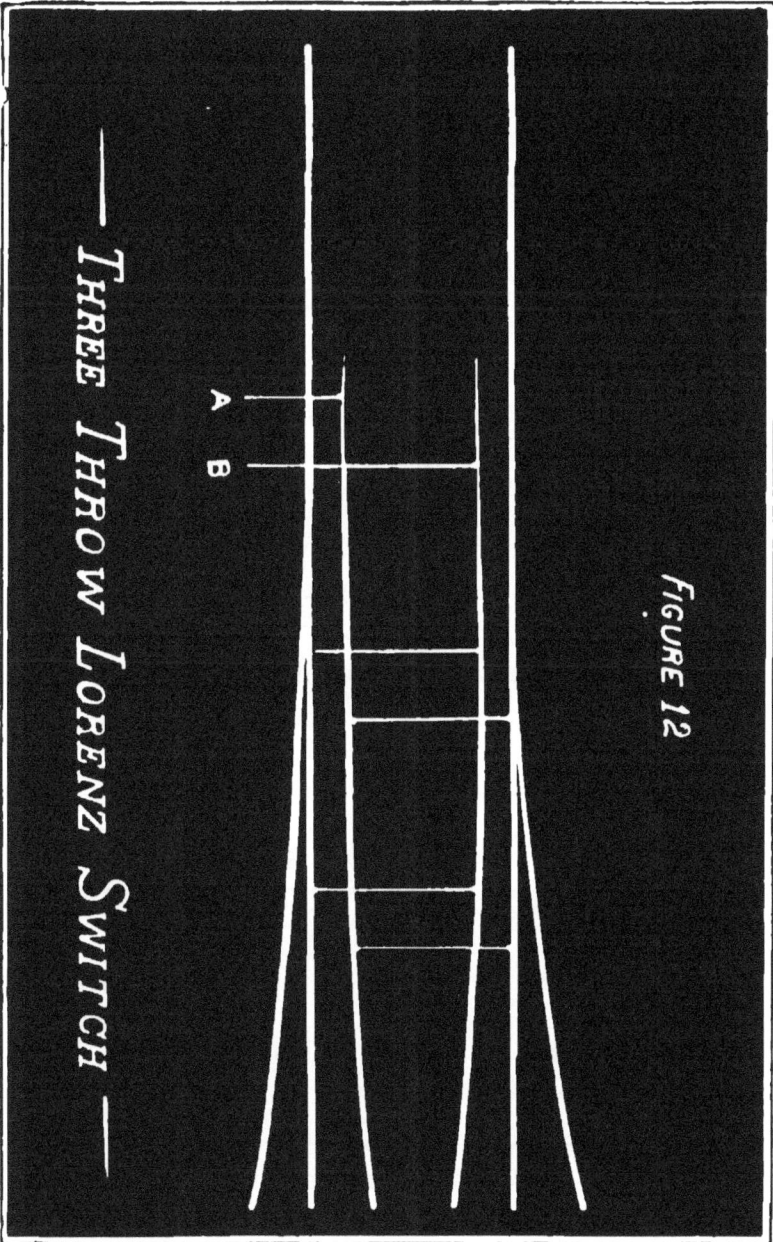

THREE THROW LORENZ SWITCH

FIGURE 12

$110 per set, including everything and ready for laying in the track. When ordering it is necessary to designate whether a right or left handed switch is wanted. The following table for correctly laying a Lorenz Switch is added for the use of Roadmasters and Engineers.

Table for Laying the Lorenz Safety Switch.
(Matthews.)

Switch Angle, 1º 30′.

Throw of Switch is the distance between flange sides of main track and switch-rails at heel of switch.

Frog Distance is the distance from heel of switch to point of frog, measured diagonally across the track.

	20 foot Switches.	24 foot Switches.	Switches. 30 foot.
Switch angle,	1º 30′	1º 30′	1º 30′
Throw of Switch,	6¼ inch	7½″	9¼″
Frog Distance,	55 feet.	53′8″	51′7″
Radius,	556.64	542.78′	521.6
Degree of Curve,	10º 18′	10º 34′	11º 00′
Mid-ordinate,	8¼ inch.	8″	7¾″

N. B.—20 foot Switch means, with point-rails 20′ long, etc.

To meet the objection raised against the Lorenz Switch that "it is dangerous to run against the points," Mr. Ainsworth, Roadmaster of the North Pennsylvania Railroad, invented and patented a switch, as shown in the figure. The switch is constructed by bending the main track rail A so as to receive the blunt pointed switch rail B when shut, which forms the lap of the rails at that point. The short or tapered switch rail C is made in the usual manner, and both are connected by the rods d d d in the usual way. The flanges of both switch rails, where they overlap the main rails, are so shaped as not to need planing, and are placed above the flange of the main rails in such a manner as to leave the switch rail flanges of their full thickness. The switch is in position for main track use when the blunt point B is placed alongside the bent main rail A. When the blunt switch or movable rail is open, it is then in position for the siding. The inner edge of the

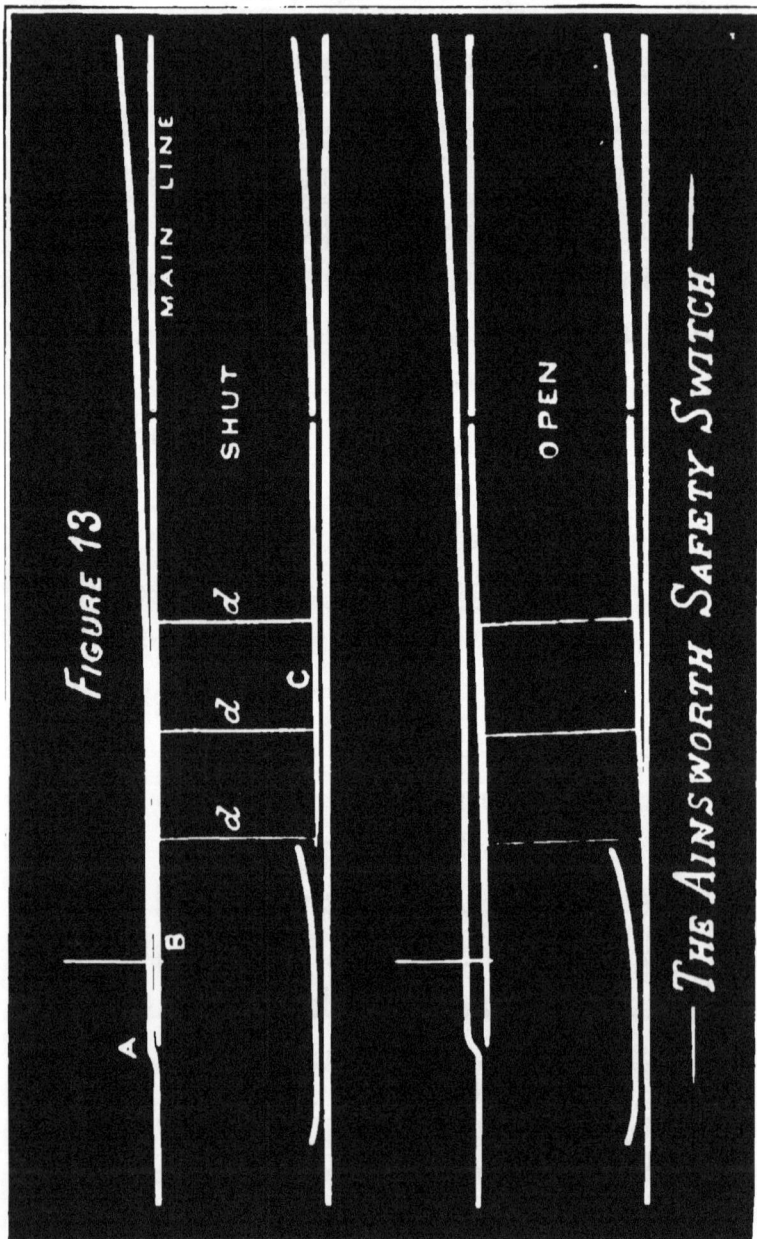

Figure 13

THE AINSWORTH SAFETY SWITCH

wheel flanges will press against the side of the switch rail which will guide them into the switch, and the distance between the flange edge of the two switch rails is so gauged that the short or tapered point is always out of the reach of the flanges of the wheels, as they are guided ' entirely by the blunt-pointed switch rail, which in consequence of its peculiar shape may be left open the space of half the width of the switch rail head without endangering the safety or direction of trains.   The automatic movement may be by spring or other known devices previously described.

The WHARTON SAFETY SWITCH, invented by Wm. Wharton, of Philadelphia, gives a main line absolutely continuous and unbroken, a decided merit, which places it far above all other switches for a road which sacrifices everything else to its main line track.   On long single track roads, like those in the West and South, where the sidings are seldom in use, and

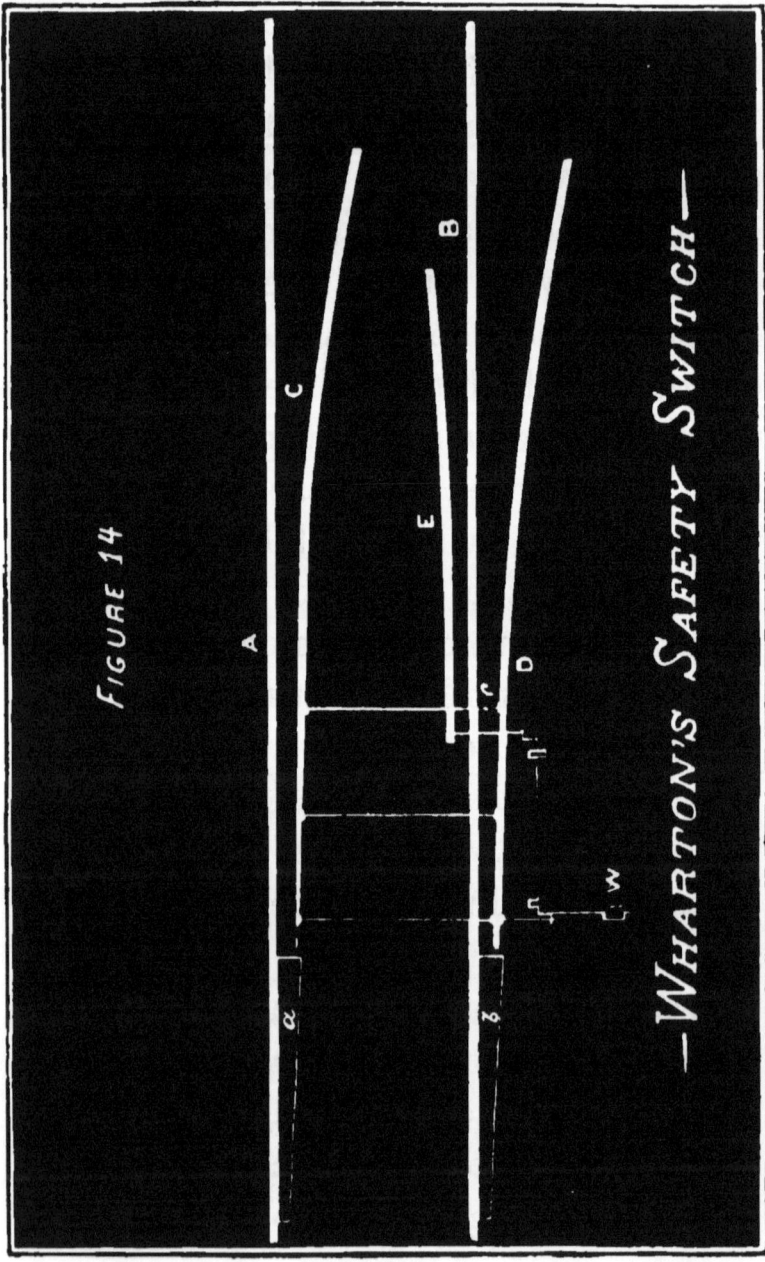

FIGURE 14

WHARTON'S SAFETY SWITCH

the main line is occupied constantly with
fast trains, this switch is preferable to any
other; but in cases where the siding is
used nearly as often as the main line, the
Lorenz Safety Switch is preferable. The
construction of this switch is rather complicated, and has many parts, but is easily
adjusted. In the diagram (fig. 14) A and
B represent the main track rails, which it
will be at once perceived are "unbroken
and continuous." C and D are two movable rails clamped together with tie rods,
and operated as shown by the lever at
W, which is supplied with a weight. C
is a grooved rail planed down to a point,
and which when thrown over against it,
fits under the head of the main rail A,
and guides the flange of one wheel out
on the siding, while the opposite wheel
gradually mounts up on the rail D, which
is somewhat higher than the main rail B,
until its flange clears the rail B, with the
tread of the wheels riding on D. The
wheel is then carried down by a gradual

decline to the proper level of the track. At the ends of the rails C D are placed two castings respectively at *b*, which, in case an engine should run *out* of the siding with the switch set for main track, receive the wheels and guide them back again to the main track. When the switch is set for main track, as shown in the figure, the curved rail E lies away from the rail B, but when set for the siding the same motion of the lever throws it over against the rail B, where it remains as long as the switch is set for the siding. Now, supposing a train to be coming on the main track from right to left, and the switch set for the siding, the first wheel flange will force the rail E away from the rail B, and consequently by means of the connecting rod *e*, will throw the lever W, and so leave the main track clear and unobstructed. The chief objection brought against this switch is, that it will not admit of fast running both ways, and is therefore exclusively a main line switch. An engine running on the side track at any very high rate of speed is very apt to rock badly while passing over the main track rails. This objection is met in a great measure by the fact that in passing in or out of a siding the train generally "slows up," and consequently passes over the switch quietly. THE COST of the switch complete is about $125 per set, ready for laying in the track.

The "SINGLE TONGUE SWITCH" is more generally known in this country as the Thiemeyer switch, owing to the fact that a Mr. Thiemeyer, of Baltimore, patented a few slight improvements on the original switch. It has been in use in Germany and in other European countries for many years. In its construction it differs from the Lorenz switch in the manner of having only one pointed tongue which is movable, the other tongue lying on the other side is a fixed tongue, making a frog. It is generally made of heavy steel castings, and although a suitable lever is much preferable, it can be used without one, the movable rail being easily moved by the foot. The switch is shown in the figure set for the main line, in which case the rail D acts as a guard rail, and, causing the wheel flanges to pass through the throat c, in a line parallel to a, keeps the wheels in the main track rails A E. Now in order to set the switch for the siding it is only necessary to close the tongue D against the main rail A, thereby necessitating the passage of the wheels over the rail D, and consequently the flanges of the opposite wheels through the throat b, past the point E, and so on to the siding rails D B. Or should the switch by accident be left set for the siding, and suppose a train on the main line passing from right to left; as soon as it reaches the rail D which, when the switch is closed, would be close up against A, the flanges of the wheels will force them apart and pass through, the

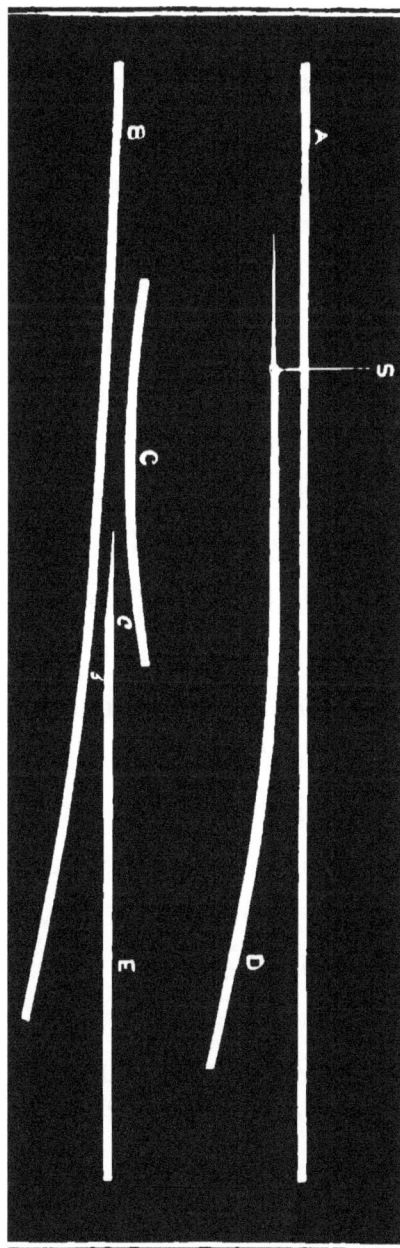

opposite rails E and B being arranged as in a frog, readily carry the other wheels from E to B; or *vice versa* if left set for main track, and a train should come down the siding. The switch works well in either case, and is a safe, reliable switch, more particularly adapted for yards, but doing good work wherever it is put, in a yard or main track. For fast running it is hardly suitable, as the continuity of the main line is broken by the frog point at E. The simplicity of its construction, coupled with the fact of its perfect safety, makes it a general favorite among railroad men. The cost of the single tongue switch complete is $90, including all attachments, per set. The Baltimore and Ohio Railroad Company have in use a great number of these switches, but have now discontinued laying them, using instead a switch invented by Mr. John L. Wilson, Roadmaster, which in design and operation is very similar to the Lorenz switch. The enormous weight of the machinery on this road necessitates very heavy track material, and a preference is given to this switch because the tongues can be made much heavier and stronger of forged steel than is possible when using a planed T-rail. The cost of making this switch is somewhat more than the single tongue and Lorenz switches, owing to the heavy forged steel tongues, which must be first hammered and then planed to shape.

The MOVABLE GUARD SWITCH is better known as WHITE'S SAFETY SWITCH, and is different from other switches, having both of the points fixed, and moving instead the two guard rails which are usually placed to guard the entrance to the switch. By shifting these guard rails from side to side the train is directed into the siding, or continues its course along the main track. Another safety switch, known as the TYLER SWITCH, is used on the Lake Shore and Michigan Southern Railroad as their standard switch. This switch is incorrectly termed the Tyler switch, as it was invented in the year 1842, by Mr. G. A. Nicolls, at that time connected with the Reading Railroad, and afterwards President of the Reading and Columbia, East Pennsylvania, and other railroads, and was in successful operation on many railroads in this country and in Cuba, when he received a patent for it in 1845. Afterwards Mr. Philos P. Tyler, of New Orleans, obtained a patent for the same device. The switch has been extensively used on the Reading Railroad, and was well known some years ago as the "NICOLLS' SAFETY SWITCH." It consists simply of two extra rails with castings at their ends so arranged that, if the switch should be left wrong, these extra rails will receive the wheels of the engine and train, and the castings will guide them safely back on to the main track. It is very economical, and could be applied to any stub switch now in use by an additional trifling cost for the extra rails and costing say $10 for each switch.

With each switch a "frog" is a necessary adjunct. The word FROG, as applied
to the railroad contrivance, is so named on
account of its supposed resemblance to
the frog of a horse's foot. Originally a
frog was nothing more than a swinging
rail pivoted in the centre, A, as shown in
the figure; but now a frog is a casting or
other mechanical contrivance to enable
the wheels of a moving train to pass over,
by, or through a point where the rail of
a siding necessarilly crosses the rail of the
main track (see fig. 16). There are two
classes of frogs, called the STIFF FROGS
and SPRING FROGS. The former denotes
a solid casting or combination of pieces
bolted stiffly together, and which when
spiked down in the track remains immovable, while the latter term implies a
frog having some of its parts arranged so
that they are adjusted by springs, and
the action of the wheel flanges wher passing over it sets it right for the tread of
the wheels to ride over. Knowing the

THE RAILWAY BUILDER.

angle with which the siding rail crosses
the main track rail, the number of the
frog to be used is known, as this angle
determines the number of the frog.
Then by referring to the preceding table
of frog distances, the distance from the
toe of the switch to where the point of
the frog should be is quickly ascertained.
Now to find the radius of the curve to
be used to connect these two points, we
have the following by Mr. Trautwine:
"From the frog angle take the switch
angle, the remainder will be the angle
at the centre of the circle; which angle
call C; subtract this angle from 180°,
and divide the remainder by 2. Call
this quotient angle A. Then as

$$\text{Nat. sine of angle C} : \text{Nat. sine of angle A} :: \text{Frog Dist.} : \text{Radius.}$$

If it is necessary to start a turnout from
a curved piece of road, the frog distance
can be found near enough for practice,
from a drawing made on a scale of about
$\frac{1}{2}$ inch to one foot. And so in the numer-

ous cases where turnouts cross tracks in various directions in and about stations, depots, etc." These can then be laid out on the ground with an ordinary tape line from the measurements given in the drawing, near enough for all purposes. Ordinary cast-iron steel plated stiff frogs of the average size (No. 8) will COST about $30 apiece. An objection to their use is the difficulty experienced in keeping them in position; being short and entirely distinct from the rails, the weight of the cars soon causes a rocking motion which tends to loosen the spikes which hold it in position. An improvement in this frog is the steel rail frog (see fig. 18), which is made up of the ordinary pattern of steel T-rails in many different styles and shapes, but in the general idea the same. In this frog the connection with the main rails is made by means of the ordinary fish plates and spikes as are used in connecting any two rails together. The PRICE of

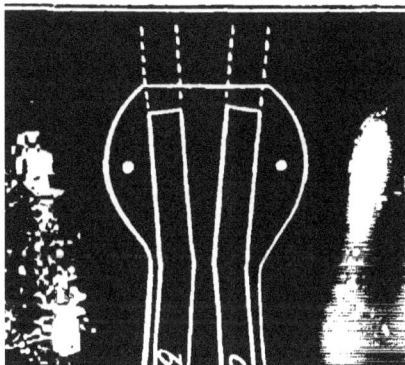

this frog (No. 8) is now as low as $28 apiece. THE SPRING FROGS are used to avoid crossing a channel or the throat of the frog, and to give a continuous even bearing to the wheels when passing over it. This is done by means of springs so fastened to the frog that the flanges of the wheels operate against the rails, and shift them into such a position that the tread of the wheel will always ride on a smooth even bearing, and will not have to cross any channel. The manner of working it is as follows: The rail A (fig. 19) and the rail C, forming together the point of the frog, are securely dovetailed together, and then riveted down to a base plate of wrought iron a. The wing rail D is also riveted to this plate sufficiently far from the point to allow the easy passage of a car wheel flange between them, usually 1¾ inches. The other wing rail B is movable, but is confined in position by rubber springs at e e, and also prevented from rising or "crawling" by a cross bar

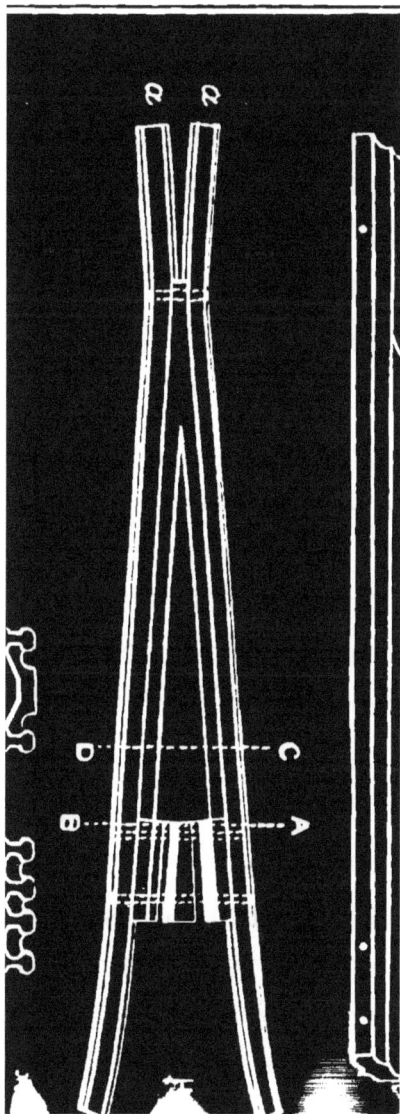

f which passes through oval slots in the two wing rails, and also the point. The frog is always kept set right for the main line by means of the rubber springs, and it will readily be observed that a train passing from A to B will do so over a smooth unbroken surface, securing much comfort to passengers, and a great saving in wear and tear to motive power and rolling stock. In passing from C to D, or on the siding, the flange of the first wheel, as soon as it reaches the wing rail B shoves it aside far enough to pass through, and each subsequent wheel in like manner, the springs allowing the necessary motion by compressing, and then (after the wheels have passed through) drawing the wing back again by their elasticity. In like manner a wheel returning from D to C will, with the aid of the guard rail on the other side of the track (which keeps the wheels "at gauge"), shove the wing rail B aside and pass through safely. These frogs are in use on the main line of the

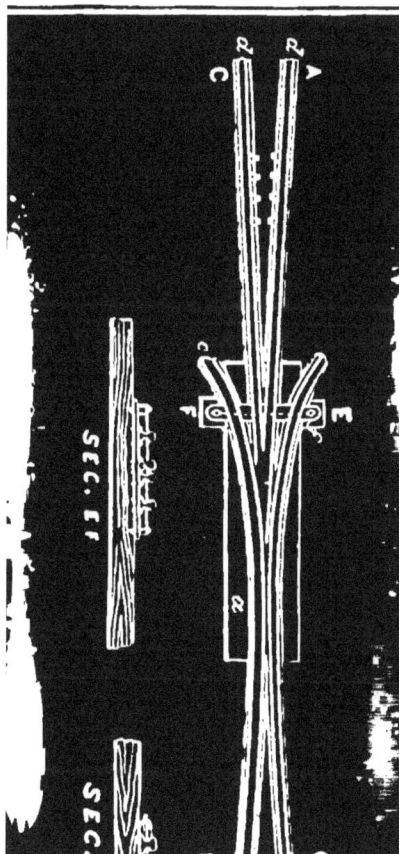

Pennsylvania Railroad and many other leading roads. They are usually made 15 feet long, and cost for an average size (No. 8) frog $50 apiece. The elastic frogs are made like stiff frogs, but have alternate layers of wood and iron, and even rubber for a base plate. The most prominent of these is the MANSFIELD FROG, which some years ago was very extensively used, but can now be hardly classed among the best frogs—the wooden base having been found to decay and crush very rapidly. The MANSFIELD FROG used to sell at $125 apiece. Among other elastic frogs, the PIERCE and the BILLINGS obtained some little favor.

A CROSSING is necessary where one railroad crosses another, and when the crossing is "at grade," that is, one railroad is on a level with the other, it is perhaps the most troublesome part of the road, and no doubt the most expensive part of the track; the former, because the utmost caution is necessary to avoid collision, and the latter on account of the double wear and tear of the rolling stock when passing over it. The danger has been obviated in several States by a law compelling all trains to come to a full stop before crossing. When it is necessary to cross a railroad, it is always advantageous to do so with a tangent and without any grade. The reason is obvious: a grade necessitating much pulling by the engine driving wheels, and consequent heavy wear on the crossing, which is liable to break or be twisted out of line. The original manner of making a crossing was simply to use rails bent to the proper shape. These were succeeded by four heavy castings heavily plated with steel, and made to fit one in each corner or intersection of the rails. Afterwards a design was used in which the rails were riveted to an iron plate which was first grooved by a planer, and strips of rubber inserted under the base of the rail, this was intended to give elasticity to the crossing. This in turn was succeeded by a CROSSING MADE

OF STEEL RAILS, firmly secured in position by bolts and fish plates in such a manner as to make the crossing virtually one piece, and gave great strength and resistance to strains in every direction. The parts of the crossing are made interchangeable, and can be easily replaced if by accident they become broken. The cost of a steel rail crossing complete, of any angle from 20° to 90° inclusive, is about $300. When laying it in position care should be taken to lay it on good heavy oak stringers, 14″×16″, or even heavier, framed to suit the angle of the crossing, and laid on a bed of broken stone at least 18″ in depth, and the ground should be carefully drained from the centre to the four corners by drains filled with broken stone, and care taken to prevent any water lodging under the stringers. Many roadmasters prefer laying crossings on sills, owing no doubt to the ease of ballasting and lining up the track, but this does not balance the good

REINFORCED POINT SWITCH.

SMALL ANGLE CROSSING.

CLAMPED, KEYED AND BOLTED STIFF FROG.

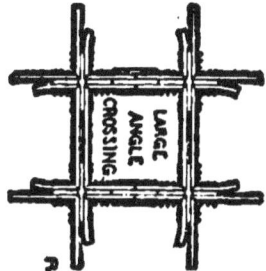

LARGE ANGLE CROSSING.

RAIL BRACE

PRESSED STEEL.

CLAMPED, KEYED AND BOLTED SPRING-FROG.

LOW AUTOMATIC SWITCH-STAND

14

result contained in a solid even bearing given by the stringers in which every inch of the rails is supported throughout their entire length. Another reason for not laying a crossing on sills is the danger of the crossing being broken in case the foundation of ballasting should give way under any one of them owing to the action of frost or of a heavy rain. Stringers will, however, stand firm even on very treacherous ground. The writer has had an extended and varied experience in constructing and laying crossings, which has only served to prove the foregoing remarks. The cost of properly laying a crossing, including the stringers, need not exceed $50. When ordering a crossing from the manufacturer the order should state the angle of intersection, section of rail to be used, gauge of both roads, and, if double track, the distance between tracks.

SIGNALS are used on railroads to notify the engineer of an engine of the position of switches, the proximity of depots or stopping places, or of any reasons why the engine should proceed or stop. Switch signals are usually made with the stand of the switches, and are called targets, the switch being altered or changed in any direction shows a corresponding change in the position of the target, which being placed at some elevation above the rails is easily seen by the engineer, and shows him how the switch is set. A FIXED SIG-NAL is a vertical post planted alongside the track, and by means of movable arms operated in a great many different ways, the engineer is notified how to proceed with his engine and train. It is not within the field covered by this work, or a detailed description of signals would be given. The subject admits of an extended description. A brief mention of the subject is made simply because calculations regarding the cost of building and operating a railroad necessarily involve signals. The arms of a fixed signal are usually painted red, blue, and white, signifying, first—*Red:* "Danger, stop!" *Blue:* "Caution, proceed slowly." *White:* "All right, go ahead." With the exception of using green for blue, this is the general interpretation of the colors. In placing fixed signals it is necessary to do so with a view to having a good background—the sky is the best—but sometimes trees or a deep cutting will prevent this, in which case an artificial background of boards painted white, or the rocks of a cutting, or the abutment of a bridge whitewashed, will tend to show the arms of a signal distinctly. On the Philadelphia and Reading Railroad SMALL TOWERS are erected for. signals at every sharp curve on the road, and at the approach to stations. These towers are placed high up in elevated positions near the track. A man is stationed in each tower, who closely watches the track which, from his elevated position, can be seen for some distance, and by turning a red, blue, or white board against the approaching train, he notifies the locomotive engineer how to proceed. At night the painted boards are replaced by colored lights.

Another class of signals, for use in foggy weather, or in case the track is obstructed at points not guarded by fixed signals, consists of small, flat, tin boxes containing powder and percussion caps. These little boxes are fastened to the rail by lead strips, and when the engine passes over them they explode with a loud report, promptly warning the engineer of "danger ahead!" The interlocking signals are too complicated to explain in detail, nor could the system be readily understood without diagrams and illustrations, which cannot be given in this work. The principle of INTERLOCKING SIGNALS is briefly defined by Barry: "If a man were to go blindfold into a signal box with an interlocking apparatus, he might, so far as accordance between points (switches) and signals is concerned, be allowed with safety to pull over any lever at random. He might doubtless delay the traffic, because he might not know which signal to lower for a particular train, but he could not lower such a signal or produce such a combination of position of points (switches) and signals as would, if the signals were obeyed, produce a collision." Saxby and Farmer's system of interlocking signals is probably the best in use. THE BLOCK SYSTEM of signals is briefly illustrated as follows :—

Suppose a line of railway to be divided off into a certain number of districts by telegraphic stations, for example, a line divided into three stations, calling the stations A, B, and C, points where three signal boxes are established. A train leaves A, and at the same time B is notified of the fact by means of the telegraph; B answers back to A, " All right, send the train," and also notifies A not to send any more trains until further orders; the

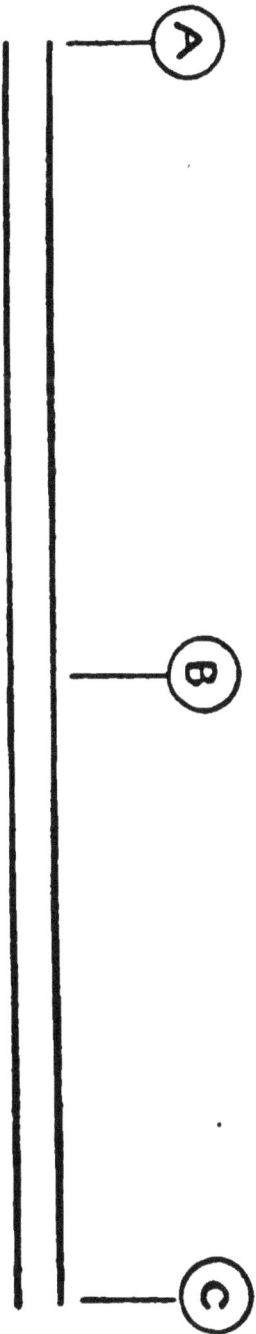

Block system of signals.

215

line is then "blocked" until the train arrives at B, that is, no other trains can run from A to B until its arrival; as soon as it arrives at B, the agent at that point notifies A of the fact, and "raises the block" between A and B, leaving the line clear. In the same manner the train proceeds from B to C, and so on through each district until it has reached its objective point. The safety and security of the block system is obvious—for supposing a train should break down or become disabled in any of the districts, say between B and C, then the "block" is continued at B, and no train can pass beyond B until the block is raised by C, which cannot be done until the disabled train has reached that point. This is the principle of the "block system," and is involved in all the improvements which have been made on it. If strictly adhered to no collision can possibly take place, as a space is always preserved between each train equal to the length of each district.

CHAPTER VI.

IN the earlier history of railways, horses were used to pull the cars which ran on tramways constructed of wood. In the year 1767, iron was substituted for wood. In 1802 Messrs. Trevithick and Vivian patented a plan for a locomotive, which ten years afterwards was put into operation. In 1811, a patent was taken out by John Blenkinsop, of Middleton, Yorkshire, England; for certain mechanical means by which the conveyance of coal and other articles was facilitated, and the expense attending the conveyance of the same was rendered less than before. It consisted in the application of a racked or toothed rail on one side of the road from end to end; into this rack a toothed wheel was worked by the steam engine, the revolutions of which produced the necessary motion without being liable to slip in descending steep inclined planes. Several of the engines were made in the years 1812 and 1813, but in the year 1814 the rack rail was abandoned, as it was found by Geo. Stevenson that the wheels adhered to the track sufficiently to do the work without it. The Liverpool and Manchester Railway was commenced in the year 1826, under the direction of Geo. Stevenson as Engineer. After mature deliberation the managment determined to have *locomotive* in preference to *fixed* engines for motive power, provided the former could be made sufficiently powerful, and that the weight were not so great as to injure the rails; also, ones that would not emit smoke. In 1829 a reward was offered for the best engine under the following conditions, viz., to consume its own smoke, to draw three times its own weight at 10 miles an hour, with not over 50 lbs. pressure of steam

The " ROCKET."

(1829.)

219

on the boiler; to have two safety valves (one locked), the boiler to be supported on springs, and to rest on six wheels if it weighed more than four and one-half tons: height to top of chimney, not over fifteen feet; weight, with water in boiler, not to exceed six tons (less preferred); boiler proved to three times the working pressure, and not to cost more than £550. This competitive trial resulted in the success of the "Rocket," and the prize of £500 was awarded to Geo. Stevenson. The superiority of the "Rocket" was due in a great measure to the boiler having small tubes, and to the blast from the exhaust steam enabling the engine to generate steam as fast as required. It is claimed that the flue boiler was suggested by Henry Booth, the treasurer of the company. From other information it appears that the tubular system of boilers was invented in both England and France at the same time. In 1835 Robert Stevenson took out a patent for leaving

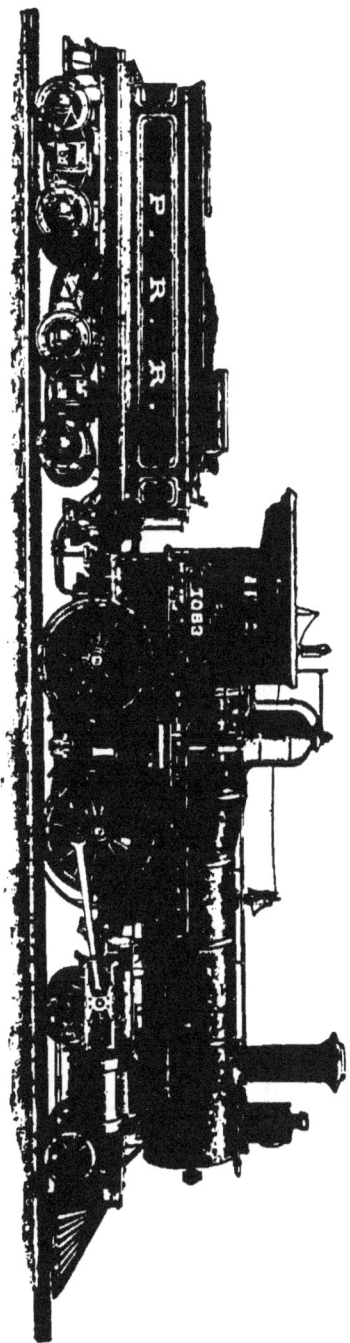

Passenger locomotive, Pennsylvania Railroad.
(1895.)

221

off the flanges of the driving wheels, and
using flanges on the leading and trail-
ing wheels only. Richard Trevithick
claims to have invented the steam blast,
but the claim has been disputed by
George Stevenson and Timothy Hack-
worth. India-rubber springs for sustain-
ing the weight of locomotives were patented by the Earl of Dundonald, in 1835.
The history of the locomotive is an
exceedingly short one, but rapid strides
have been made towards perfection in
its construction, and particularly by
American manufacturers. The total
number of locomotives in use in this
country in the year 1895 was about
37,000, and this great motive power
has been created within a few years.
The subject is an important one, and
every railroad man should be familiar
with the lever which operates his road.
The cost of a first-class passenger locomotive of the " American" pattern, as

built by the Baldwin Locomotive Works of Philadelphia, ranges from $7000 to $8000, supposing the road to be of the ordinary gauge of 4' 8½". The following dimensions of this style of locomotive are reliable.

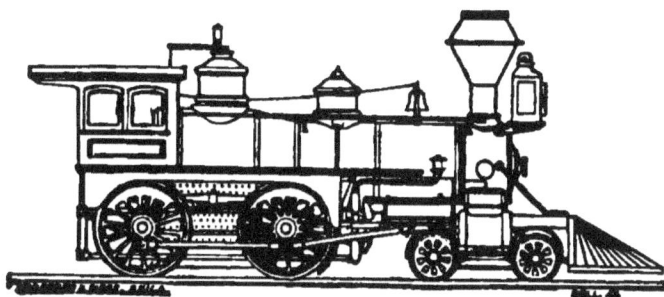

"AMERICA."

Number of driving wheels, . . 4
    "   front truck wheels . . 4
    "   back "   " . . None.
Total wheel base . . . . 21' 9"
Between centres of front and back driving wheels . . . . . 96 inches.
Total weight of locomotive, working order . . . . . . 65,000 lbs.
Total weight on driving wheels . . 42,000 lbs.
Diameter of driving wheels . . 60¾ inches.
    "   truck " . . 28 "
    "   cylinders . . . 16 "

Stroke of cylinders	24 inches.
Outside diameter of smallest boiler ring	48 "
Size of grate	$65'' \times 34\frac{1}{2}''$
Number of tubes	144
Diameter of tubes	2 inches.
Length of tubes	10 ft. 11 in.
Square feet of grate surface . .	15.5
" " heating surface in fire box . . .	100.6
" " heating surface in tubes	825.4
Total feet of heating surface . .	926.0
Exhaust nozzles (single or double) .	Double.
Diameter of nozzles	$2\frac{3}{4}$–$3\frac{1}{4}$ in.
Size of steam ports	$1\frac{1}{4} \times 15$ "
" exhaust ports . . .	$2\frac{1}{2} \times 15$ "
Throw of eccentrics	$5\frac{1}{2}$ inches.
Outside lap of valve	$\frac{3}{4}$ inch.
Inside " "	$\frac{1}{32}$ "
Size of main driving axle journal .	$7''$ dia. $\times 8''$
" other " " " .	$7''$ dia. $\times 8''$
" truck axle journal . . .	$4\frac{1}{2} \times 7\frac{1}{2}$ in.
Diameter of pump plunger . .	2 inches.
Stroke of pump plunger . . .	24 "
Capacity of tank	2000 gall.

The style known as the "Mogul" is used for freight purposes, and its cost ranges from $8000 to $9000.

The following are the principal dimensions:—

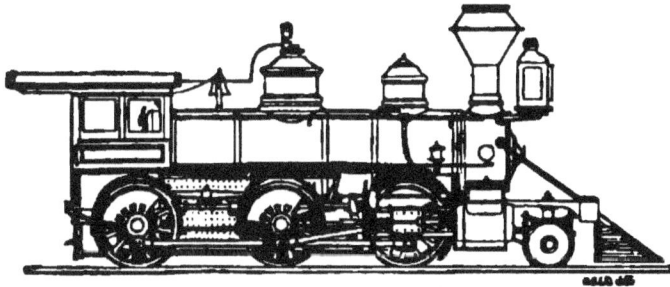

EQUIPMENT.

"MOGUL."

Gauge of road	4′ 8½″
Number of driving wheels . . .	6
" front truck wheels . .	2
Total wheel base	22′ 8″
Distance between centres of front and back driving wheels . . .	96 inches.
Total weight of locomotive . . .	77,000 lbs.
" on driving wheels . .	66,000 "
Diameter of driving wheels . .	52 inches.
" of truck wheels . . .	30 "
" of cylinders . . .	18 "
Stroke of cylinders	24 "
Outside diameter of smallest boiler ring	50 "
Size of grate	66″×34½″
Number of tubes	161
Diameter of tubes	2 inches.
Length of tubes	11′ 3″
Square feet of grate surface . .	16′ 0
" " of heating surface in fire box	102.7

15

Square feet of heating surface in tubes	948.0
Total square feet of heating surface .	1051.0
Exhaust nozzles	Double.
Diameter of nozzles	3″ to 3½″
Size of steam ports	1¼×16″
" exhaust ports . . .	2½×16″
Throw of eccentrics	5½ inches.
Outside lap of valve	¾ "
Inside " "	1/32 "
Size of main driving axle journal .	7″ to 8″
" other " " " .	7″ to 8″
" truck " " " .	5″ to 8″
" pump plunger . .	2 inches.
Stroke of pump plunger . . .	24 "
Capacity of tank	2200 galls.

The third style of engine, also a freight engine, is called the "Consolidation." The following dimensions of this class of locomotives is from the Danforth Locomotive Works, Paterson, New Jersey. The cost of this locomotive will range from $9000 to $10,000 for standard (4′ 8½″) gauge.

"CONSOLIDATION."

Gauge of road	4' 8½''
Number of driving wheels . . .	8
" front truck wheels . .	2
Total wheel base	23' 2''
Distance between centres of front and	
back driving wheels . . .	15' 7''
Total weight of locomotive . . .	96,550 lbs.
" " on driving wheels . .	86,430 "
Diameter of driving wheels . .	4' 2''
" truck wheels . . .	2' 7''
" cylinders . . .	20''
Stroke of cylinders	24''
Outside diameter of smallest boiler	
ring	4' 2''
Size of grate	120''×34¾''
Number of tubes	165
Diameter of tubes	2¼''
Length of tubes	13' 9¼''
Square feet of grate surface . .	29
" " heating surface, fire box	139
" " " tubes	1370

Total feet of heating surface . .	1509
Exhaust nozzle	Double.
Diameter of nozzle	$3\frac{1}{2}''$
Size of steam ports	$1\frac{1}{2}'' \times 15\frac{1}{2}''$
" exhaust port	$2\frac{3}{4}'' \times 15\frac{1}{2}''$
Throw of eccentrics	$5\frac{1}{3}''$
Outside lap of valve	$\frac{5}{8}''$
Inside " "	None.
Size of main axle journal . . .	$6\frac{3}{4}''$
" other driving axle journal .	$6\frac{3}{4}''$
" truck axle journal . . .	$5''$
Diameter of pump plunger . .	$2\frac{1}{4}''$
Stroke of pump plunger . . .	$24''$
Capacity of tank	2400 galls.

The " Bicycle" engine used on the Philadelphia and Reading Railway has the following dimensions:

Gauge	$4' 8\frac{1}{2}''$
Diameter of high-pressure cylinders .	$13''$
" low " " .	$22''$
Stroke of high-pressure cylinders .	$26''$
" low " " .	$26''$
Diameter of driving wheels . .	$84\frac{1}{4}''$
Wheel base of engine . . .	$22' 9''$
Weight of engine in working order .	115,000 lbs.
" on driving wheels . .	48,000 "
Diameter of boiler	$56''$
Length of fire box, Wootten pattern .	$114''$

Passenger locomotive "Bicycle," Philadelphia and Reading Railway.

Width of fire box, Wootten pattern .	96″
Heating surface of combustion chamber	45 sq. ft.
Heating surface of fire box . .	128 "
" " tubes . . .	1293 "
Total heating surface . . .	1466 "
Capacity of tender	4000 galls.

Regarding the other dimensions, the *height of the stack* is usually from 13 to 15 feet above the rails. The *tenders* of engines will weigh about 6 tons empty, and about 15 tons full of water and fuel. Width of tender 9 feet, and length about the same as the engine. The price given for the locomotive is supposed to include the tender.

The size of American locomotives has steadily increased with no apparent limit in view. The decapod engines, with 10 driving wheels, weigh as much as 148,-000 pounds,—these are freight engines. On the Erie Railroad are some eight-wheeled passenger locomotives weighing 115,000 pounds. The approximate price of any locomotive of the usual standard sizes may be estimated at about 8 cents per pound weight.

THE FIRST RAILWAY CAR ever used for carrying passengers was built in 1825, and was run on the Stockton & Darlington Railroad, in England. It was simply a common box-car made of wood, with three windows on each side, and mounted on four fixed wheels. The English cars, even at the present day, vary but little from this design; but the American cars are constructed on an entirely different principle, having two swinging or bogie trucks under each end. AN AMERICAN PASSENGER CAR measures about 50 feet in length from out to out of bumpers, with an extreme width of about 10 feet. The trucks have four, six, and sometimes eight wheels each. A car will hold 60 passengers comfortably, and when empty will weigh about 15 tons, when filled with passengers, about 19 tons.

Cost of a Pennsylvania Railroad Passenger Car.

The following table gives in detail the cost of constructing one first-class standard passenger car, at the Altoona shops of the Pennsylvania Railroad, the total cost being $4423.75. The principal items are as follows:—

Labor	$1263 94
Proportion of fuel and stores . . .	28 61
2480 feet poplar	86 80
3434 feet ash	127 08
1100 feet pine	20 90
2350 feet of yellow pine	70 50
500 feet oak	10 00
450 feet hickory	13 50
700 feet Michigan pine	49 00
400 feet cherry	16 00
439 feet maple veneer	24 14
4 pairs wheels and axles . . .	332 85
2 pairs passenger car trucks, complete	533 62
13 gallons varnish	52 34
45 lbs. glue	14 33
2925 lbs. iron	87 75
792 lbs. castings	16 99
Screws	51 88

Passenger car, Pennsylvania Railroad.

Gas regulator and gauge . . .	25	25
2 two-light chandeliers . . .	50	72
2 gas tanks	84	00
1 air-brake, complete. . . .	131	79
57 sash balances	44	61
61 lights glass	65	83
2 stoves	77	56
25 sets seat fixtures	50	50
3 bronze lamps	13	50
2 bronze door locks and fittings . .	15	20
Butts and hinges	15	58
13 basket racks	77	35
12 sash levers	42	00
61 bronze window lifts . . .	24	40
61 window fasteners	16	47
238 sheets tin	41	44
273 lbs. galvanized iron . . .	25	31
96 yards scarlet plush. . . .	228	87
44 yards green plush	109	99
61 yards sheeting	10	30
243 lbs. hair	72	95
12 springs	22	96
12 spiral elliptic springs . . .	20	29
1 head lining	80	63
2 packets gold leaf	14	58
Various small items	261	44
	$4423	75

At present there are about 28,000
passenger cars in use on the railways

American sleeping car.

234

of the United States, and about 8000 of baggage, mail, and express cars.

SLEEPING CARS are about the same size as ordinary ones, but some have been constructed 70 feet long by 11 feet in width, and weighing about 33 tons when empty. Ordinary passenger cars cost from $4000 to $5000 apiece, and sleeping cars from $6000 to $20,000 apiece. Other "special" cars are used for various purposes. A modern special train can now be found on nearly every trunk-line in America. It consists of cars having every convenience, including a composite car, with electrical dynamo and engine, a barber-shop, and bath-room; a dining-, sleeping-, drawing-room, library, and observation car, —all vestibuled together and practically continuous. A mail or baggage car of about the same dimensions as a passenger car will cost from $1500 to $2500 each. FREIGHT BOX-CARS measuring 30 feet in length and 9 feet in width, with

8 wheels, will weigh about 8 tons, and cost about $500 each. Platform or GONDOLA CARS of the same dimensions weigh about 7 tons, and cost about $350 each.

The total number of freight cars in use on our railways is now about 1,230,000.

Average Weight of Cars.

4 ft. 8½ in. gauge, as in use on principal railways in the United States.

	Weight.
Baggage car, 36 feet out to out,	28,000 lbs.
Mail " " " "	32,000 "
Passenger " 48 " "	37,000 to 39,000 "
Sleeping " " " "	41,000 to 44,000 "
Stock " " " "	17,500 to 18,500 "
Box, " " " "	16,400 to 17,800 "
Flat " 32 feet long,	16,500 "
8-wheeled coal cars,	13,440 "
4 " "	6,720 "
Iron trucks for passenger cars,	16,000 "
Wooden trucks for passenger cars,	14,000 "
Passenger car axles, of iron,	330 "
' " steel,	320 "
Freight " iron,	300 "

A freight car will carry from 15 to 25 tons of weight. Coal cars usually carry

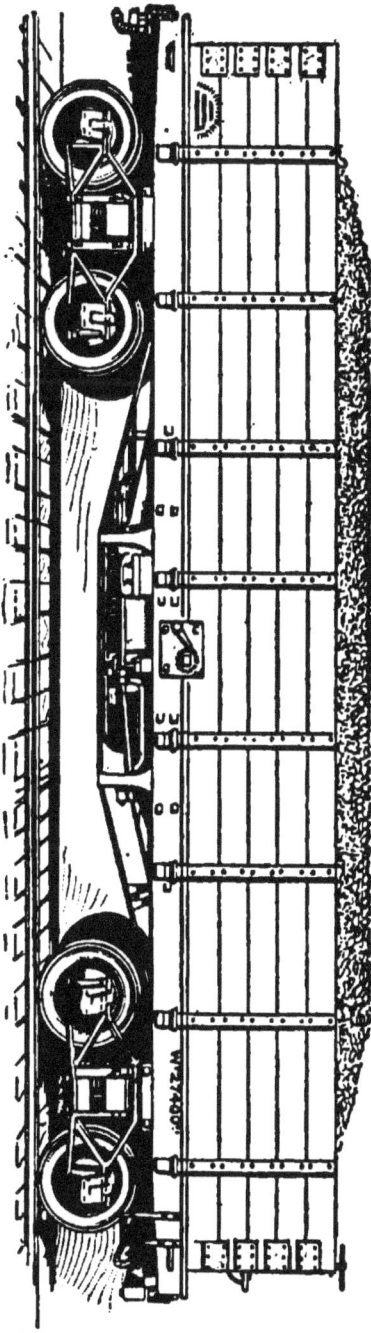

American coal car.
(30 tons capacity.)

about 5 tons of coal for the smaller, and 25 for the larger cars.

Weights of Substances.

Salt	℔ bushel	80	lbs.
"	barrel	280	"
Cement	"	300	"
Water	gallon	8.33	"
Oil	"	7¾	"
Alcohol	"	6.96	"
Turpentine (spirits) . .	"	7.31	"
Powder	keg	25	"
Coal, anthracite . . .	bushel	86	"
" bituminous . .	"	80	"
Coke	"	32	"
Charcoal (hard) . . .	"	30	"
Asphaltum	c. foot	56	"
Gutta percha . . .	"	61	"
India-rubber . . .	"	60	"
Ivory	"	112	"
Pitch	"	71	"
Rosin	"	68	"
Tar	"	62	"
Sand	"	120	"
Slate	"	175	"

When the weights of articles of freight are not given, the above weights are established by railroad agents.

A cubic yard of sand			weighs about 30 cwt.
"	"	gravel	" " 30 "
"	"	mud	" " 25 "
"	"	sandstone	" " 39 "
"	"	shale	" " 40 "
"	"	granite	" " 42 "
"	"	slate	" " 43 "

COAL CARS having only four wheels weigh about 3 tons, measure about 12 feet long by 6 feet wide, cost about $180 each, and will carry about 5 tons of coal. The maximum load for the foregoing freight cars, with eight wheels, is 30 tons. Recently these cars have been constructed of steel, weighing about 40,000 pounds, with a capacity of 100,000 pounds of coal. THE WHEELS for passenger and freight cars are usually from 30 to 33 inches in diameter; lately larger diameters have been strongly advocated, and wheels have been made measuring 42 inches in diameter. In America the car wheels are usually made of cast-iron in one piece, and the tire is united to the hub with a disk or plate. Some are made with single, and others with double plates, and still others with spokes similar to a driving-wheel. The tread of the wheel is hardened by a process called *chilling*. This is done by pouring the melted cast-iron into a mould of the form of the tread of the wheel. The mould is also made of cast-iron, but being cold cools the melted iron very suddenly, and thus hardens it somewhat as steel is hardened when it is heated and plunged into cold water. Only certain kinds of cast-iron will harden in this way—the cause is not known.

The manufacture of these wheels is an industry requiring long experience and nice judgment; not only as to the character of the irons used, but also regarding their combination, so as to secure all the qualities required in the product.

All kinds of chilled cast car wheels are embraced in the following forms, viz.: double plate, single plate, and spoke,

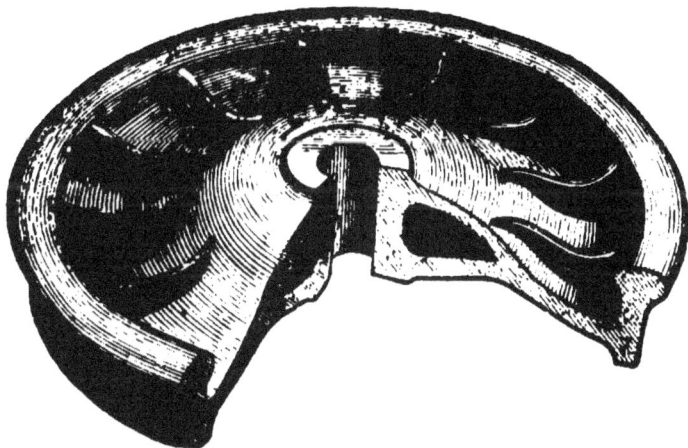

Section of double-plate wheel.

which are varied in numerous ways to carry out the views of the makers. While much depends on the form of the pattern for securing an even depth of chill on the tread, freedom from strain, and a maximum strength for the weight of iron used, much more depends upon the proper selection and manipulation of the iron to be used. Charcoal pig iron, possessing the greatest strength and best chilling properties, is exclusively used by all first class manufacturers of wheels; while the anthracite irons, some of which also chill, are used to a large extent by makers of inferior wheels. The wearing surface or tread of the wheel is chilled and made exceedingly hard by running the molten metal into iron moulds which have been previously turned to the required form. To make a good wearing and even chill is the great desideratum. A chill of moderate depth, say $\frac{1}{2}$ of an inch to $\frac{5}{8}$ of an inch, is sufficient for service, and, if tough and hard, will wear better than double that depth. A chill should be bright and finely grained, or fibrous looking, gradually merging into the gray iron back of it, and, if solid and hard, the wheel will often stand a service of 100,000 miles wear. Some kinds of charcoal iron possess great strength but with a limited chilling property, while in others the chill is very marked, but in coarse and ragged crystals. In fact very few charcoal irons can be found which will, by themselves, make a good wheel. For this reason all wheel founders use a mixture of different irons in order to get the required qualities of the several kinds and grades of iron combined in the wheel, but in nearly every case the mixture is simply made in charging the cupola, and, as some iron melts more rapidly than others, the desired result is not obtained. At the foundry of a chilled-wheel works the several kinds of iron are not only thoroughly mixed in the cupola, but care is also taken that those irons which melt easily are charged so as to melt at the same time as the more difficult. Then all iron is drawn from the cupola into large reservoirs, holding from 15 to 25 tons each, and thoroughly puddled and mixed before and during the pouring of the wheels, and, being poured at the highest possible temperature, results in smoothness of tread and a tough chill of even depth, having also strength of plate and freedom from strain. These latter qualities are further increased by annealing for four days in air and moisture-proof pits, made of iron cylinders let 10 feet into the ground, and lined with fire brick. The wheels are then taken from the pit and cleaned with wire brushes, after which they are carefully inspected, and are ready for sale.

Cast-iron wheels weigh as follows :—

Spoke	wheel,	14″ dia.	3″ tread,	90 lbs.
"	"	16 "	3 "	120 "
Single plate "		18 "	$3\frac{1}{2}$ "	150 "
" " "		20 "	$3\frac{3}{4}$ "	200 "
" " "		24 "	$3\frac{3}{4}$ "	250 "

Double plate wheel,	26″ dia.	4¾″ tread,	375 lbs.
" " "	28 "	4¾ "	415 "
" " "	30 "	4¾ "	500 "
" " "	33 "	4¾ "	525 "

For tenders and
passenger cars.

" " "	33 dia.	4¾ "	560 "
" " "	42 "	4¾ "	875 "

CAR WHEELS ARE WORTH about 2.0 cents per pound of weight, and will run about an average of 50,000 miles before wearing out; although cast-iron wheels have shown much longer service, often twice as much mileage. About 70,000 miles, however, is considered a good mileage for a cast-iron wheel. Cast-iron wheels with STEEL TIRES are often used for passenger cars and locomotive tenders and trucks. They possess an advantage from the fact that, after the tread of the wheel becomes hollowed out by wear, it can be turned off on a lathe to a true bearing. An elastic steel-tired wheel, invented by Anson Atwood, is constructed with hemp packing inserted between the body of the wheel and the tire, and possesses considerable merit. One of these wheels, measuring 33″ diameter, was run under a Wagner sleeping-car somewhat over 250,000 miles, during two years' service, in which time the tire was turned off twice, losing one-half of an inch of its diameter. A steel-tired wheel is worth about 5 cents per pound.

Table of Steel-tired Wheels.

33″ diameter,	4″ tread,	weighs	659	lbs.
33 "	4¾ "	"	700	"
30 "	4 "	"	585	"
30 "	4¾ "	"	615	'

The car, engine truck, and driving AXLES require especial attention in purchasing—quality being of greater account than price. As made at the best works, they meet all the requirements of railroads. They are hammered from selected scrap wrought-iron, which is first mechanically cleaned of rust, thereby securing a more thorough

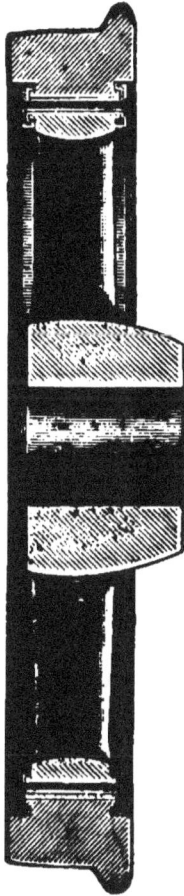

Section of steel-tired spoke wheel.

247

welding in the subsequent operation of
shingling, faggoting, and swedging into
form; the journals being hammered in
to secure a denser and harder bearing
surface. Driving axles are used from 5
inches to 8 inches in diameter; engine
truck axles, from $4\frac{1}{4}$ to $5\frac{1}{2}$ inches diameter; and locomotive tender and passen-
ger and freight car axles, from 4 to 5
inches diameter, and having a length of
from 6' 6'' to 8 feet. The concave form
of axles, which can only be obtained by
hammering, is preferable, for, by distri-
buting the vibrations, the shoulder of
hub seat is relieved from the liability
to fracture. Axles made of old rails are
very inferior in quality, and, consequently,
can be sold very cheap—at about 2 cents
per pound—while the best kind of hammered axles, made from selected scrap,
are worth about $2\frac{1}{2}$ cents per pound. The
cores of the wheels are reamed out by
revolving table drills, and the journals
and wheel seats of the axles are also

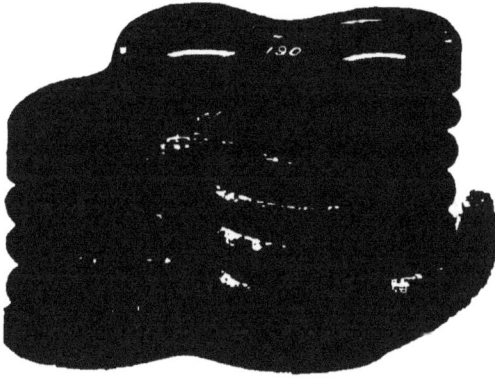

Bolster car springs.

249

carefully turned off on a lathe, after which the wheels are pressed on to the axles by a hydraulic machine at a pressure of about 30 tons weight.

WROUGHT-IRON FRAMES for the trucks of cars will weigh about 175 pounds each, and cost about 5 cents per pound, finished and fitted together. There are two frames to each truck, these two constitute a "set." COUPLINGS of wrought iron are sold by the pound, and vary in price according to the market. SPRINGS are made of the best crucible cast-steel, and in various forms and shapes. Springs made of round steel, and of spiral shape, are much used, and cost as follows:—

Buffer or draw springs, 11 cents per lb.

Bolster springs, in sets of 4 springs, $20 to $30 per set.

Equalizing bars springs, in sets of 8 springs, $30 to $35 per set.

If it be required to arrest the speed of a train, the BRAKES are applied, and the train is stopped. Various kinds of brakes have been in use both in this country and abroad, differing from each other in the construction, but generally alike in regard to the kind of power employed to operate them. One kind of brake is known as the SLIPPER BRAKE, and consists of a contrivance by means of which friction is produced between the brake block, or slipper, and the rails. Another brake is applied directly to the wheels, as in an ordinary road wagon. Still another kind is a clip which is attached to the car, and by means of screws and levers is made to grip the rail firmly between the arms of the clip. Again, there are brakes operating directly on the axle, instead of the tread of the wheel; but the usual manner of operating brakes is by the strength of the brakeman's arm operating a set of levers, which causes the brake blocks to press close up against the tread of the wheels, and so stops their motion. This manner of braking injures the wheels to a great extent, sometimes by locking the wheels so that they remain immovable, and slide along the rails, making flat places on their periphery. On passsenger trains, however, the AUTOMATIC OR CONTINUOUS BRAKES are now extensively used, which are so arranged that the engine driver can apply the brakes of all the cars simultaneously; they are also so constructed that should the cars by any means become uncoupled, the same action will immediately apply the brakes on the detached car, and so arrest its motion. THE WEST-INGHOUSE BRAKE is constructed on this principle, and so is the LOUGHRIDGE. The power used is simply compressed air, which is conveyed by pipes from the receiver, which is attached to the locomotive, to every car, at which point it enters a cylinder, and, operating on a piston and levers connected with it, so applies the brakes. The air is compressed by means of a steam pump on the engine, and the receiver is a receptacle for the compressed air until it is required to use it. Experiments made in England with the Westinghouse automatic brake resulted as follows: A train of thirteen carriages, travelling at the rate of 52 miles an hour, was brought to a full stop in *19 seconds!* The distance run from the time the brakes were applied until the train stopped was 913 feet. The time formerly required to stop a train in England by hand brakes, train running at the rate of 49½ miles per hour, was from 60 to 90 seconds, and in a distance of from 2000 to 4000 feet.

LOUGHRIDGE claims the best stop yet made, in a trial on the Baltimore and Ohio Railroad, in which a train of ten cars travelling at the rate of $42\frac{6}{10}$ miles per hour was stopped in 16 seconds and after running only 587 feet 8 inches. The cost of equipping a train with these brakes is comparatively trifling The railroad companies usually pay the inventor for the right to use the brake per mile of road, or per car, or whatever other form of royalty they choose to pay, and then make the brake apparatus, and fit it to their cars in their own shops. Other brakes operated by water, or by creating a vacuum, are in occasional use, but are inferior to either of the above. SMITH's VACUUM BRAKE, however, showed very good results in a competitive trial—Clark's "hydraulic" also.

The time is not far distant when all freight, as well as passenger, cars will be equipped with the air-brake. Laws have been suggested compelling all railroads to adopt this very practical and useful invention.

CHAPTER VII.

DEPOTS AND STRUCTURES.

IT is highly important when building a railroad to consider well what depot facilities are *necessary* to accommodate the business of the road, and not how much money can be wasted in a ponderous unsightly pile of brick and stone, oftentimes immeasurably inconsistent with the business to be transacted therein. PASSENGER STATIONS should be made with a view to the comfort of passengers. The tracks upon which the trains arrive and leave should be systematically arranged for "outward bound" and "incoming" trains, with separate and distinct platforms and exits for each class of passengers, either those arriving or departing. A covering should always be provided amply large enough to cover the standing trains, both arriving and departing; under this same covering, sufficient offices for baggage, waiting-rooms for ladies and gentlemen, and a well-ordered restaurant should be provided. The cost of such a depot will, of course, vary considerably, so much in fact that no very near estimate can be given. The value of the property not being taken into consideration, however, the building itself need not cost more than $10,000. While first operating the road, more attention should be given to the improvement of the roadbed than to the depots, but at all times there is no excuse for not making the passengers comfortable while waiting for arriving or departing trains.

In the construction of a FREIGHT DEPOT the most important feature is to consider the easiest possible way of landing and transferring the freight which is to pass through it. All known mechanical contrivances for facilitating the handling of the particular kind of freight which the

road is intended to carry should be carefully examined by the Engineer, and the
merits of each carefully weighed in his
mind with a view of adopting whatever
may be of value. Sufficient siding room
should be provided for empty cars waiting to be loaded, and these sidings should
be arranged with a grade descending in
the direction in which the cars are to
move, so that the men employed around
the building can easily move them without
the expense of a shifting engine. All these
little details must be thought of and provided, and every Engineer will have to
study out his own plan of a freight depot
to suit his own particular case. WAY
STATIONS, as a rule, need only be small
buildings sufficiently large to protect the
passengers from the weather, and also of
sufficient capacity to store away a limited
amount of freight. It is quite common
to have a building for a way station so
arranged that one-half of it can be occupied by the station agent and his family,

17

leaving the other half for depot purposes; such a building need not cost more than from $3000 to $5000, according to the style in which it is built. FLAG STATIONS are small settlements where little or no freight business is done, and where the trains do not stop unless they are signalled by a flag or other signal to do so. They are simply wooden shelters from the weather, often having only three sides inclosed, the fourth being left open. Their cost should be about $100. A fixed signal is usually placed at a convenient distance from the station, so that it can be operated by the passenger wishing to board the train, providing there is no station agent at that point, in which case the brakemen of the train lower the signal again before the train moves off. A TURNTABLE is a revolving platform for turning locomotives around horizontally. Suppose, for example, a straight piece of track with the terminal ends A and B: the engine starts from A and pulls the train to B; now, in order to run the en-

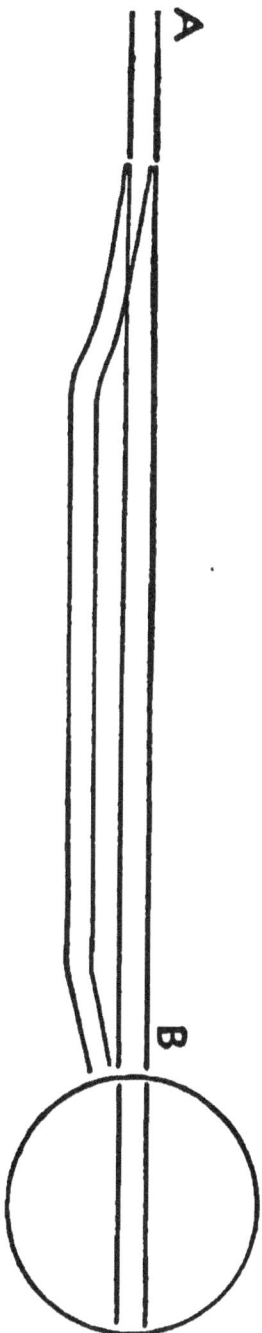

Railroad turntable.

A

B

gine back to A, it must run backwards or else be turned around by some mechanical means; for this purpose a turntable is used. Turntables are made of wood or iron and in every variety of construction. A wooden one is not so expensive as iron, but is not so desirable to have. A good wrought-iron one of 50 feet diameter, to carry an engine and tender weighing 133,000 pounds, will cost from $900 to $1000; while one 45 feet in diameter will cost from $730 to $800 complete. WATER STATIONS are placed at certain points along the line of the road to supply the engines with water. They consist of small frame structures surmounted by a tank with a capacity of from 5000 to 50,000 gallons of water. These tanks are pumped full of water by steam pumps, hand power, horse power, and in the West very often windmills are used for this purpose. The tanks are built circular in shape or often square, and are used as reservoirs for water, being always kept full. The water

Cast-iron water column.

may be supplied from a neighboring spring, brook, river, or a well can be dug and the water drawn from that to fill the tank with. The cost of a circular tank capable of holding about 25,000 gallons of water will be about $300 put into position. Steam is generally used for roads doing a large business. A good steam pump with boiler attachments complete, ready for work, can be estimated at about $500. A cast-iron stand which is connected with pipes to the reservoir or tank is placed alongside the rails, and in such a manner that a canvas hose conveys the water from it into the tank of the locomotive; the flow of water being regulated by a lever operated by the fireman of the locomotive. Many of the fast trains on the Pennsylvania R. R. do not stop to take water, but from a long tank placed between the rails, which is kept filled with water, the Engineer, letting down a scoop from the locomotive, scoops up the water while the train is in motion. Regarding the FUEL for locomotives, the reports are very conflicting as to the superiority of one kind over another. Wood, bituminous coal, and anthracite coal are used, and each is strongly advocated by the road using that particular kind. On the Philadelphia and Reading Railroad, a passenger engine drawing a train of five cars consumes about 1½ tons of anthracite coal in 100 miles running. The freight engines use about 3½ tons in the same distance, and the heavy coal engines consume about 4½ tons. On the Baltimore and Ohio Railroad a passenger engine drawing five coaches consumes only 1 ton of Cumberland bituminous coal in a distance of 100 miles, and the freight engines consume about 3 tons. On the Pennsylvania Railroad the passenger engines consume about 1½ tons of bituminous coal every 100 miles of "run" made. Wood burning engines will consume about 2½ cords of wood every 100 miles. A cord of wood is 4 × 4 × 8 feet, or 128 cubic feet. A ton of anthracite coal is 40 cubic feet.

Properties of Fuel.

Kind of fuel.	Lbs. of water evaporated per pound.	Per cent. of carbon.	Cubic feet of air required for 1lb. of coal.	Weight per cubic foot.	Cubic feet to stow a ton.
Bituminous coal	7 to 9	80	265	50	44
Anthracite	8 to 10	92	282	54	40
Coke....	8 to 10	86	245	31	72
" Natural Virginia	8 to 9	80	260	48	48
" Cumberland	8 to 10	80	250	32	70
Charcoal.................	5 to 6	96	265	24	104
Dry wood	4 to 5	44	147	20	100
Wood, 20 p. ct. water	4	34	115	25	100
Turf, dry (peat)	6	51	165	28	80
Turf, 20 per ct. water	5	40	132	30	75
Illuminating gas......	13.8	...	194	0.37	29,800
Oil, wax, tallow	14	71	200	59	37
Alcohol	9.56	58	154	52	42

Stationary engines use 3 to 7 pounds of coal per horse-power per hour.

Locomotive passenger engines 25 to 30 pounds coal per mile.

Locomotive freight engines, 45 to 55 pounds coal per mile.

Wood-burning locomotives, 1 cord of wood to 40 miles.

Bulk of wood about 6 times as much as an equivalent of coal.

At certain points on a long road, or at the terminal station of a short road, a COALING PLATFORM must be erected in order to supply the engines with coal. These platforms are so arranged that coal can be conveniently hauled on to them and stored there; when an engine is to be supplied, the coal is dropped through shutes from the platform into the tender of the locomotive. The cost of these platforms will depend on their size.

ENGINE HOUSES are built of wood or brick, depending entirely upon the necessities of the road. A house built in the form of a segment of a circle is the most convenient form for a road operating a large number of locomotives. A turntable is generally placed at the entrance to an engine house both for convenience in "housing" the engine and to turn them around whenever required. A brick engine house of this kind, designed for a number of engines, will cost about $1000 per engine stall,—a wooden house is only intended as a temporary covering, and is very seldom used.

ROAD CROSSINGS, or points along the line of the railroad where it crosses common wagon roads and streets, are, as a rule, grade crossings, that is, the street is on the same level as the railroad which crosses it. On some well regulated railroads, running through crowded cities and crossing streets at grade, gates made of iron are put up and so arranged that no wagons can cross the railroad until the train has passed. This arrangement is also used in England at every grade crossing, and every precaution is taken to avoid accidents to passing vehicles. There are many situations in which a bridge to pass a highway over a railway track would be very expensive, and this is a strong reason for having level crossings; and in some situations, near stations, there would be inconvenience in the approaches to the railway. That highways should pass over or under a railway no one will doubt. Level crossings have caused much personal injury, the loss of many lives, persons maimed for life, and much damage to property. The locating engineer should consider well every facility for obviating the necessity of a level crossing, especially where there may be obstructions preventing a good range of sight. The subject is of sufficient importance to justify some modifications in the grade or alignment, if by that means a highway may be passed by a bridge. To raise an embankment to the elevation required to pass over the railroad sometimes interferes with other avenues or objects to such a degree that it cannot well be done; but if there be nothing of this kind, and the expense of raising the bank for landing the highway over the railway be the only objection, the case should be very unfavorable to warrant a level crossing. The objection to merely an elevation of the highway, as injuring the public accommodation for travel, is by no means worthy of being placed against the benefits the travellers on the public road will derive from the safety from collision that will be secured by the bridge.[1] The same reasoning holds good as to farm crossings. Serious accidents have occurred on these when made over the level of the rail. A good timber farm bridge, crossing a single track road, can be built for about $500, and the damages arising from one collision would pay the entire cost.

[1] Jervis, "Railway Property."

CHAPTER VIII.

THE miles of Railway built in the United States probably exceeds that in all the rest of the world.

This development has taken place within the past fifty years, and is remarkable for the enormous amount of invested capital. The average statistical mind finds it difficult to comprehend such figures as $11,300,000,000, and that amount represents the money put into American Railways up to the present writing. Beginning with only 23 miles in 1830, this mileage has grown to 181,000 miles in 1895. Every year of the intervening time has witnessed the steady addition to the previous figures.

In 1887 there was over 12,000 miles of new railway built, the greatest number of miles for any single year. 1882

269

was the next best year for new railways
in the United States, some 11,000 miles
having been built during that period.
The following table shows the progress made each year :

*Table showing the Number of Miles of
Railways in the United States.*
(Poor.)

Year.	Miles.	Year.	Miles.	Year.	Miles.
1830	23	1852	12,908	1874	72,385
1831	95	1853	15,360	1875	74,096
1832	229	1854	16,720	1876	76,808
1833	380	1855	18,374	1877	79,088
1834	633	1856	22,016	1878	81,767
1835	1,098	1857	24,503	1879	86,584
1836	1,273	1858	26,968	1880	93,296
1837	1,497	1859	28,789	1881	103,143
1838	1,913	1860	30,626	1882	114,712
1839	2,302	1861	31,286	1883	121,455
1840	2,818	1862	32,120	1884	125,379
1841	3,535	1863	33,170	1885	128,361
1842	4,026	1864	33,908	1886	136,379
1843	4,185	1865	35,085	1887	149,257
1844	4,377	1866	36,801	1888	156,169
1845	4,633	1867	39,250	1889	161,353
1846	4,930	1868	42,229	1890	166,698
1847	5,598	1869	46,844	1891	170,769
1848	5,996	1870	52,922	1892	175,188
1849	7,365	1871	60,293	1893	177,465
1850	9,021	1872	66,171	1894	179,393
1851	10,982	1873	70,268	1895	181,021

In 1895, according to Poor's Manual, THE TOTAL INVESTMENT in Railways in the United States, including their capital stock, funded and unfunded debt, was $11,362,985,080. Their NET EARNINGS during this same year was $327,505,716, or only 2.9 per cent. on capital as compared with 4.7 per cent. in the year 1880.

The COST PER MILE of road as measured by the amount of their stocks and bonded indebtedness equalled $60,188.

Their GROSS EARNINGS amounted to $1,105,284,267, or 9.7 per cent. on invested capital as against 11.36 per cent. in 1880,—facts that need no comment.

The General Exhibit of all Railroads in the United States for the year 1895 :

Liabilities.

Capital Stock..............................	$5,182,121,999
Funded Debt...... ....…...	5,640,942,567
Unfunded Debt......	418,505,092
Current Debt.............................	429,331,956
Total Liabilities................	$11,670,901,614

Assets.

Cost, Railroad and Equipment......	$9,861,102,973
Real Estate, Stocks, Bonds, etc	1,683,909,608
Other Assets...............................	259,804,963
Current Accounts........	224,706,821
Total Assets.....................	$12,029,524,365

Net Assets...............$358,622,751

INDEX.

18   273

Contract, form of, 56.

Corps, constructing, 84; of engineers, 15.

Cost of car wheels, 245; cast-iron frogs, 200; culvert masonry, 162; passenger locomotives, 222; switch chairs, 169; trestling, 153.

Cost per mile of railways, 271.

Couplings, wrought-iron, 250.

Crossings, at grade, 206; how to lay, 210; of steel rails, 208; road, 266.

Cross-over, 163

Cross section, of railroad cutting, 87; of tunnel, 98; work, 85.

Cross ties, cost of, 109; Burnettizing, 108; number to a mile, 109; on American railroads, 107; sawed, 108.

Cubic yard of earth, 91; of rock, 92.

Cubical contents, 85.

Culvert, foundations for, 162; how built, 161; masonry, 63.

Cumberland and Pennsylvania Railroad grades, 81.

Curvature of the earth, 42.

Curves, 39.

D.

Depot and structures, 255; freight, 256; passenger, 255.

Dooley's stub switch, 174.

Double throw switch, 172.

Drains, double box, 161; on a curve, 162; on a skew, 162.

N.

Narrow gauge spikes, 136.
Nicolls, safety switch, 196; single joint switch, 175.
Nuts and bolts, 129; weight of, 131.

O.

Obstacles to survey, 235.
Open joint, 122.
Outfit for field party, 23.

P.

Passenger locomotive, 221; station, 255.
Pennsylvania Railroad ballast, 104; fish plate, 126; passenger car, 231.
Permanent way, 102.
Philadelphia and Reading Railroad grade, 79; tunnels, 100.
Pick, work of a, 86.
Pierce frog, 206.
Plough, work of a, 86.
Pocket level, 24.
Points, running against the, 182.
Port Clinton tunnel, 100.
Preliminary line, 23; survey, 45; survey, cost of, 49.
Prismoidal formula, 85, 94.
Profile of railway, 83.
Proposals, contract, 56; railroad, 76.

R.

S.